TRIAZENES
Chemical, Biological, and Clinical Aspects

TRIAZENES
Chemical, Biological, and Clinical Aspects

Edited by

Tullio Giraldi
University of Trieste
Trieste, Italy

Thomas A. Connors
Medical Research Council
Toxicology Unit, Carshalton
Surrey, United Kingdom

and

Giuseppe Cartei
Division of Medical Oncology
Ospedale S Maria della Misericordia
Udine, Italy

SPRINGER SCIENCE+BUSINESS MEDIA, LLC

Library of Congress Cataloging-in-Publication Data

International Conference on Triazenes: Chemical, Biological, and
  Clinical Aspects (1989 : Trieste, Italy)
    Triazenes : chemical, biological, and clinical aspects / edited by
  Tullio Giraldi, Thomas A. Connors, and Giuseppe Cartei.
       p.   cm.
    "Proceedings of an International Conference on Triazenes:
  Chemical, Biological, and Clinical Aspects, held November 28-29,
  1989, in Trieste, Italy"--T.p. verso.
    Includes bibliographical references and index.
    ISBN 978-1-4613-6710-9   ISBN 978-1-4615-3832-5 (eBook)
    DOI 10.1007/978-1-4615-3832-5
    1. Triazenes--Therapeutic use--Testing--Congresses.
  2. Antineoplastic agents--Congresses.   I. Giraldi, Tullio.
  II. Connors, Thomas A., 1934-   III. Cartei, Giuseppe.
  IV. Title.
    [DNLM: 1. Triazenes--congresses.   QD 305.A9 I61t 1989]
  RC271.T68I58   1989
  616.99'4061--dc20
  DNLM/DLC
  for Library of Congress   90-14271
                                                       CIP

Proceedings of an International Conference on Triazenes: Chemical,
Biological, and Clinical Aspects, held November 28 – 29, 1989,
in Trieste, Italy

ISBN 978-1-4613-6710-9

© 1990 Springer Science+Business Media New York
Originally published by Plenum Press, New York in 1990
Softcover reprint of the hardcover 1st edition 1990
All rights reserved

INTRODUCTION

More than 25 years have elapsed since the development of the seminal idea which led to the synthesis of dimethyltriazenes as antitumor agents. The original suggestion of Shealy et al. was to use 4-imidazone-carboxamide as the carrier of a nitrogen-containing cytotoxic function. 5-diazoimidazole-4-carboxamide (diazo-IC) was synthesized and tested in mice as a potential inhibitor of de novo purine biosynthesis. Its lack of antitumor action was attributed to its polarity and to the resulting poor uptake of this hydrophilic chemical. Diazo-IC was then coupled with dimethylamine, yielding 5, (3,3-dimethyl-1-triazeno)imidazole-4-carboxamide) (DTIC) with the intention of obtaining a less polar and more lipophilic prodrug which might release diazo-IC intracellularly. Preliminary tests showed that DTIC had good antiumor activity in experimental systems. Further tests demonstrated a broad spectrum of action against rodent tumors, and clinical trials indicated activity against human malignancies. Subsequent clinical use of DTIC has demonstrated its usefulness against malignant melanoma, for which it is currently the drug of choice, and its effectiveness in combination chemotherapy in the treatment of other human cancers. Because of its antitumor activity the mechanism of action of DTIC has been investigated in some detail. The original rationale for its development, that is, the hydrolysis in vivo to diazo-IC, has been shown not to be involved in the mechansims of action. DTIC requires metabolic activation before it exerts its biological effects. Biotransformation of DTIC is via hepatic oxidative N-demethylation, forming the corresponding monomethy derivative MIC [5-(5-methyl-1-triazeno)imidazole-4-carboxamide], which is a reactive metabolite methylating cellular components. Other products of biotransformation have been identified, particularly for aryl-dimethyltrazenes, which may also be relevant to their in vivo effects. This pharmacological evidence may explain other biological properties of DTIC such as its mutagenic effects in both prokaryotic and eukaryotic cells. Also in tumor-bearing animals it appears to stimulate immune responses as a result of chemical xenogenization of the tumor cells. It may also modify the malignant properties of tumor cells markedly, reducing their metastatic potentials in a relatively stable fashion.

These biological and pharmacological properties are not unique for imidazole dimethyltriazene but are also seen with aryldimethyltriazene, which permits an easier and more detailed examination of structure-activity relationships. In this connection, of a large series of benzenoid derivatives CB10-277 [1-p-(3,3-dimethyl-1-triazeno)benzoic acid sodium salt] has interesting properties and is currently on clinical trial in England. Another approach arising for studies of DTIC analogues has led to the synthesis of chemicals with cyclic moieties, such as mitozolomide and temozolomide, which can yield methyl- and β-chloroethyltriazene <u>in vivo</u>.

This volume contains contributions by the participants in the "International Conference on Triazenes: Chemical, Biological, and Clinical Aspects," which was held in Trieste on November 28-29, 1989. Synthesis, chemical properties, biotransformation, and pharmacokinetics in animals and humans are reported in the following chapters. Similarly, antitumor effects in animals and man are illustrated, together with other biological and pharmacological properties. Studies on imidazotetrazines and antitumor melamines, such as hexa-and penta-methylmelamine, are included because of the analogies of these compounds to triazenes. One chapter reports the work of a workshop session devoted to the alkylation of O^6-guanine in the mechanism of action of triazenes, and another summarizes the main aspects of the poster session.

The Editors wish to express their appreciation to the contributors to this volume, and to all of the participants in the Conference. Special acknowledgments are due to the following agencies which supported the meeting, thus making this volume possible: Università di Trieste, Regione Autonoma Friuli - Venezia Giulia, and Consiglio Nazionale delle Ricerche. The patronage of Società Italiana di Farmacologia (SIF), Associazione Italiana di Oncologia Medica (AIOM), and Società Italiana di Cancerologia (SIC) is also gratefully acknowledged. The Editors wish to express particular recognition of the invaluable contribution provided by Dr. Sonia Zorzet during the organization of the Conference and the preparation of this volume.

Tullio Giraldi
Thomas A. Connors
Giuseppe Cartei

CONTENTS

TRIAZENES: SYNTHESIS AND CHEMICAL PROPERTIES

Keith Vaughan

Department of Chemistry, Saint Mary's

University, Halifax, N.S., Canada   B3H 3C3

INTRODUCTION

It is a great pleasure to participate in this, the first international conference brought together to discuss the Chemical, Biological and Clinical Aspects of Triazenes. This paper deals mainly with the chemistry of triazenes; the biological properties of these compounds will be discussed in detail in other chapters of these proceedings. However, where appropriate, the anti-tumour activity of relevant compounds will be described briefly here.

Interest in the chemistry of triazenes has been sustained largely by the significant in vivo anti-tumour activity of the 3,3-dimethyltriazenes, $ArN=N-NMe_2$. It is important to describe briefly the known metabolism of these triazenes in order to appreciate the design of chemical experiments aimed at the development of new triazenes with improved activity.

It is generally accepted that 3,3-dimethyltriazenes require metabolic activation to a cytotoxic species and that the course of the metabolism involves oxidation to give the hydroxymethyltriazene, $ArN=N-NMe-CH_2OH$ (I) [Scheme 1]. Loss of formaldehyde from I affords the "monomethyltriazene", $ArN=N-NHMe$ (II), which is presumed to be the active metabolite due to its proclivity for methylation of DNA. The thrust of this paper is to review the chemistry of the triazenes of type I and II, which have been synthesised and scrutinised in this laboratory during the last ten years.

Triazenes, Edited by T. Giraldi *et al.*
Plenum Press, New York, 1990

Scheme I Triazene metabolism

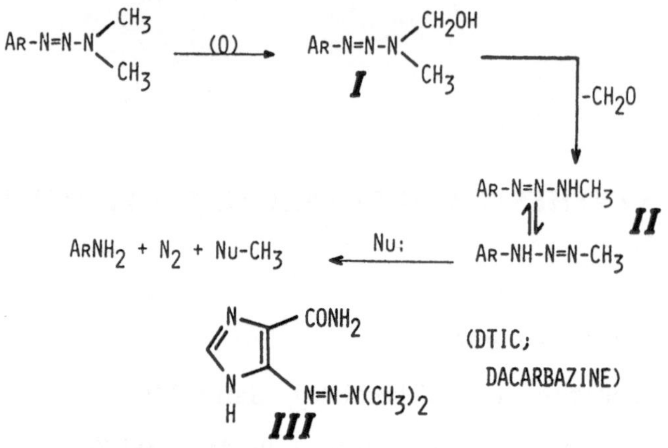

$$Ar-N=N-N\diagdown\!\!\!\begin{matrix}CH_3\\CH_3\end{matrix} \xrightarrow{(O)} Ar-N=N-N\diagdown\!\!\!\begin{matrix}CH_2OH\\CH_3\end{matrix}$$

I

$$\downarrow -CH_2O$$

$$Ar-N=N-NHCH_3$$
$$\updownarrow \qquad II$$
$$ArNH_2 + N_2 + Nu-CH_3 \xleftarrow{Nu:} Ar-NH-N=N-CH_3$$

(DTIC;
DACARBAZINE)

III

MONOMETHYLTRIAZENES

The chemistry of the monoalkyltriazenes, ArN=N-NHR, has been reviewed (1). Diazonium coupling of the arene diazonium salt with a primary aliphatic amine (equation i) is a convenient source of the monoalkyltriazene:

(i) $ArN_2^+ + RNH_2 \longrightarrow ArN=N-NHR + H^+$

However, this apparently simple approach can often lead to higher nitrogen homologues, the pentaazadienes, ArN=N•NR•N=NAr. The formation of the monoalkyltriazene in reaction (i) is favoured by the presence of electron withdrawing substituents in Ar (2).

Monoalkyltriazenes are not stable in aqueous media and undergo degradation to give a mixture of triazene and non-triazene products (3). The principal products of the hydrolysis of II are the arylamine, $ArNH_2$, the diaryltriazene, ArN=N•NHAr, and the 1,3-diaryl-3-alkyltriazene, ArN=N•NRAr. These products are indicative of the tautomeric nature of the monoalkyltriazene:

ArN=N-NHR $\rightleftharpoons$ ArNH-N=N-R
    IIa   IIb

Nuclear magnetic resonance studies of these triazenes confirm the tautomeric behaviour in solution and, furthermore, show that the position of the equilibrium is influenced by the nature of substitutents in the aryl moiety (4). Thus 3-methyl-1-p-tolyltriazene in deuteriochloroform exists almost exclusively as the "conjugated" tautomer (IIa); the chemical shift of the methyl group in this tautomer is close to the N-methyl chemical shift of a 3,3-dimethyl-triazene and at low temperature the N-methyl proton resonance of IIa becomes a doublet due to spin-spin coupling to the adjacent NH proton. In contrast, the NMR spectrum of 3-methyl-1-p-cyano-phenyltriazene shows a very broad room temperature N-methyl signal which resolves into a singlet (for IIb) and a doublet (for IIa); apparently the presence of an electron-withdrawing substituent in Ar shifts the equilibrium towards the "unconjugated" tautomer (IIb). Quite a novel efect is seen when a nitro-group is positioned ortho- to the triazene moiety; in this case, only the presence of IIb is indicated by a singlet methyl resonance in the NMR. Presumably hydrogen bonding between the nitro and the NH groups locks the molecule into the "unfavourable" tautomeric form.

The tautomeric behaviour of monoalkyltriazenes in solution would suggest that the biological activity might be influenced by the facility of nucleophilic displacement at tautomer IIb. However, there is no evidence to suggest that the anti-tumour activity of the 3,3-dimethyltriazene is affected by the electronic character of the substituent in the 1-aryl moiety.

The requirement for metabolic activation of the dimethyltriazene may be a factor in limiting the clinical efficacy of these drugs. It has been suggested (5) that metabolism of N-alkyl compounds is species dependent. The plasma half-life of the dimethyltriazene, DTIC (III), is significantly greater in the rat than in the mouse, and greater still in man. The poor clinical activity of DTIC may be due at least in part to the low level of metabolism in man. A possible solution to this problem is to devise a pro-drug for DTIC, one that avoids the requirement for oxidative metabolism. A possible pro-drug candidate is a monomethyltriazene; however, the poor stability of monomethyltriazenes in aqueous solution has precluded such application.

An alternative candidate for a pro-drug is the hydroxymethyltriazene, or one of its derivatives, and the remainder of this paper will explore the chemistry of the hydroxymethyltriazenes and their known derivatives.

HYDROXYMETHYLTRIAZENES

Until relatively recently, these triazenes were thought to exist only as transient metabolic intermediates and that their chemical isolation would not be feasible. It turns out that, under appropriate conditions, the synthesis of a hydroxymethyltriazene is straight-forward. Diazonium coupling with an aqueous mixture of methyl-amine and formaldehyde (equation ii) affords the crystalline hydroxymethyltriazene, which once isolated is a stable crystalline solid (6).

(ii) $ArN_2^+ + MeNH_2 / CH_2O \longrightarrow ArN=N-N\begin{smallmatrix} CH_2OH \\ \\ Me \end{smallmatrix}$

The "appropriate" conditions for success of this approach are (a) a large excess of formaldehyde over methylamine, and (b) the presence of a strongly electron withdrawing group in Ar. Other diazonium salts react with methylamine/formaldehyde to give a mixture of triazenes; in addition to the hydroxymethyltriazene, a second type of triazene with a higher molecular weight is formed. This latter product has been unequivocally identified as the N,N-bis(1-aryl-3-methyltriazen-3-ylmethyl)methylamine (IV):

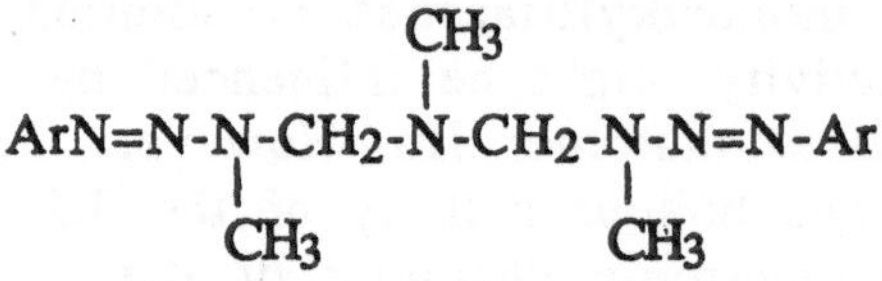

$$\underset{\underset{CH_3}{|}}{ArN=N-N}-CH_2-\underset{\overset{CH_3}{|}}{N}-CH_2-\underset{\underset{CH_3}{|}}{N}-N=N-Ar \qquad (IV)$$

These "bis-triazenes" were identified coincidentally by our group (7) and by Iley and co-workers (8). Their formation is favoured by (a) the absence of -M groups in Ar, and (b) a low formaldehyde/methylamine ratio. A rational explanation for the formation of the two types of triazene is provided by the complex set of equilibria suggested in Scheme 2. Condensation of formaldehyde with the alkylamine gives the carbinolamine, $RNHCH_2OH$; direct diazonium coupling with this species gives the hydroxymethyltriazene whereas further condensation with the alkylamine can give a number of polymeric amines. These polymers can couple with diazonium ions to give the observed products, IV and also V, which was isolated in one case only by coupling with p-chlorobenzene diazonium chloride (9).

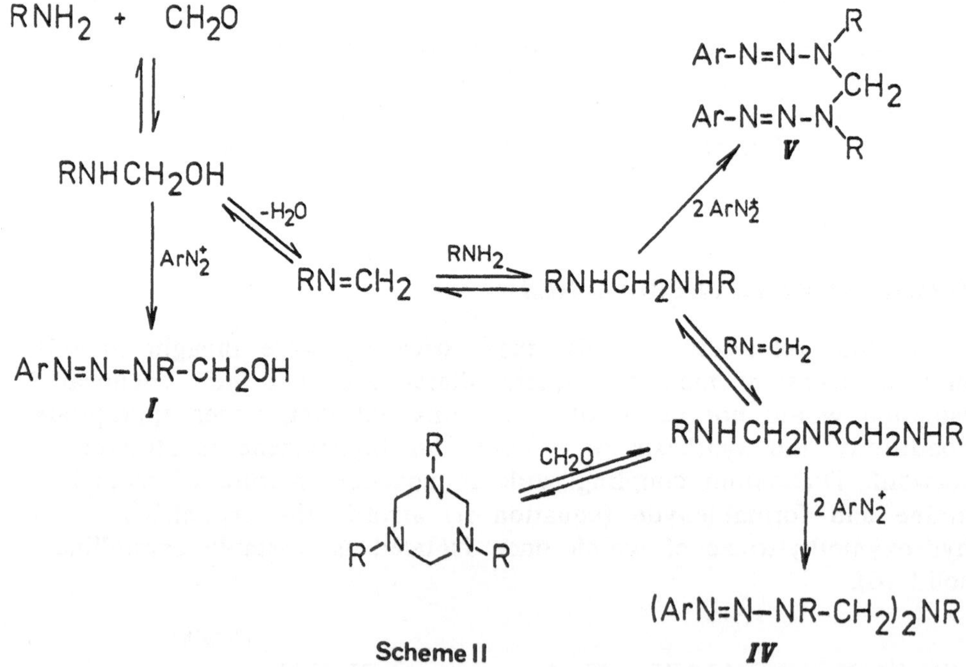

Scheme II

4

The hydroxymethyltriazene structure (I) has been confirmed by IR, NMR and mass spectral analysis (10) and also be X-ray diffraction (11). The mass spectra of the hydroxymethyltriazenes suggest that they fragment by loss of formaldehyde to give the monomethyltriazene, which is also the product of hydrolysis in solution. The hydrolysis of a hydroxymethyltriazene in aqueous buffer has been followed by UV spectroscopy, and the half-life is very similar to that of the corresponding monomethyltriazene. The hydroxymethyltriazenes have pronounced anti-tumour activity against the TLX5 tumour in vivo (10); in vivo - in vitro bioassay experiments suggest that the hydroxymethyltriazenes exert their in vivo anti-tumour activity via the degradation product, the monoalkyltriazene.

Hydroxymethyltriazenes behave generally as "masked" forms of the monomethyltriazenes; the chemical and biological properties of the two series of compounds are parallel. For example, a series of alkyltriazenyl imidazoles have been investigated for their differential cytotoxicity towards the HT-29 (Mer+) and BE (Mer-) cell lines and for their ability to cause DNA strand breaks and cross-links. A monomethyltriazene, and a hydroxymethyltriazene capable of generating the monomethyltriazene in situ, were preferentially cytotoxic towards the BE cell line compared with the HT-29 cell line (12). The BE cell line is deficient in the repair of O_6-methylguanine lesions.

Several crystalline derivatives of the hydroxymethyltriazene have been observed. Reaction of the HMT with acetic anhydride in pyridine affords the acetate (VI) and with benzoyl chloride in pyridine the benzoate (VII) (13). Both types of ester, VI and VII, undergo solvolysis in methanol to give the methyl ether (VIII), and a kinetic study of this and other solvolysis reactions (14) has adduced evidence for the $S_N{}^1$ iminium ion mechanism shown in Scheme 3:

Scheme III

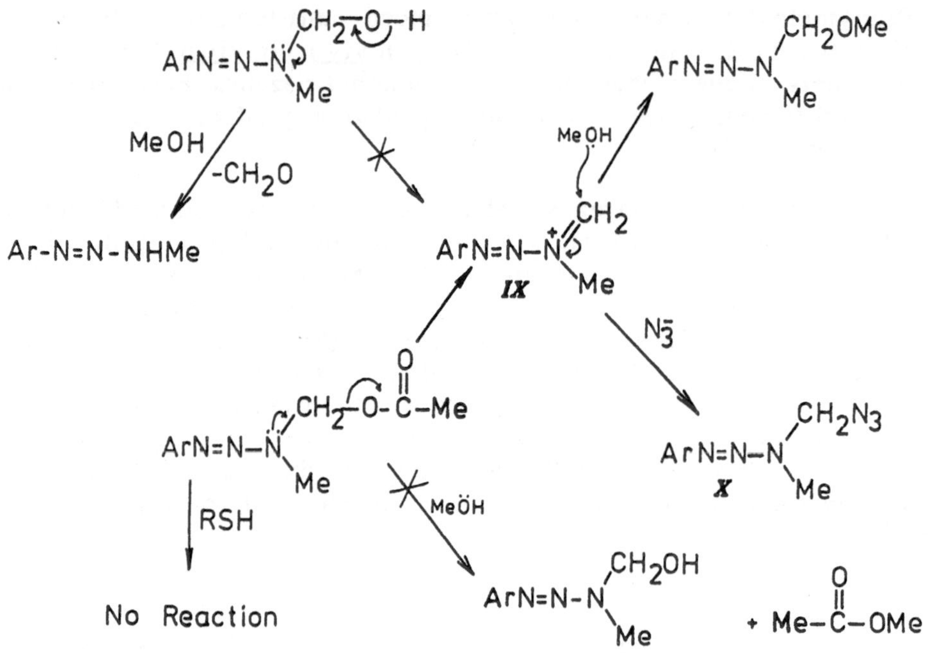

Solvolysis of the HMT itself in methanol results only in loss of formaldehyde and there is no evidence for iminium ion formation directly from the HMT. However, the rate of solvolysis of the acetate (VI) has been correlated with the Grunwald-Winstein parameter for solvent ionising power, thus supporting the hypothesis of an $S_N{}^1$ mechanism. The iminium ion (IX) is then captured by methanol to give the observed ether and can be trapped by azide ion to give the novel azidomethyltriazenes (X) (15), which have shown anti-tumour activity against the P388 and PC6 tumours.

The acetoxymethyl- and azidomethyltriazenes (VI and X) appear to hydrolyse rapidly in buffer and behave biologically like "masked" hydroxymethyltriazenes. They display selective toxicity towards the BE cell line which is deficient in the repair of O^6-methylguanine lesions. Moreover, recent studies of the kinetic parameters of the azides suggests that they have half-lives significantly longer than the other derivatives so far described and may have significant potential as pro-drugs.

The recognition of iminium ions as intermediates in these reactions suggests that an electrophilic intermediate may be involved in the metabolism of a dimethyltriazene. It is not inconceivable that the hydroxymethyltriazene could be conjugated in some way, e.g., a glucuronate, that would facilitate loss of the α-leaving group just as acetate appears to be lost from VI. The hypothesis that reactive, electrophilic iminium ions may be generated by the metabolism of a variety of N-alkyl amines is not new and several such observations have been reported (16).

The alkylamine/formaldehyde/diazonium salt recipe has been put to good use in the synthesis of some novel heterocyclic systems. Thus the reaction of arenediazonium ions with a mixture of formaldehyde and ethanolamine affords spontaneously a new series of 3-(arylazo)-1,3-oxazolidines (XI) (17). Although these oxazolidines look structurally like precursors of the hydroxymethyltriazene (XII), none of the compounds in this series appear to undergo ring-opening in aqueous solution. However, the method has been extended to the synthesis of some tricyclic heterocycles, e.g., the oxazolidinobenzo-triazine (XIII), by incorporating the carbonyl moiety of the formaldehyde component in the ortho-position of the arene-diazonium ion.

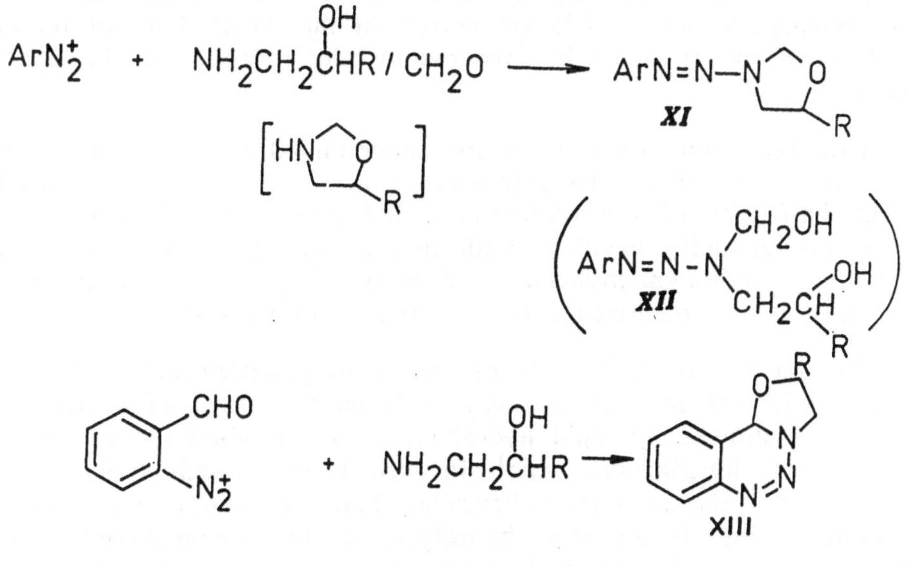

$$ArN_2^+ \ + \ NH_2CH_2\overset{\overset{\displaystyle OH}{|}}{C}HR \ / \ CH_2O \ \longrightarrow \ ArN=N-N\overset{\frown}{}O$$

XI

$$\left[HN\overset{\frown}{}O \overset{}{\underset{R}{}} \right]$$

$$\left(ArN=N-N\overset{CH_2OH}{\underset{CH_2\overset{\overset{\displaystyle OH}{|}}{C}H}{}} \right)$$

XII

$$\begin{array}{c}\text{CHO} \\ \\ -N_2^+\end{array} \ + \ NH_2CH_2\overset{\overset{\displaystyle OH}{|}}{C}HR \ \longrightarrow \ \text{XIII}$$

Another variation on the original recipe was to replace methylamine with ammonia in order to provide a route to the synthesis of the monohydroxymethyltriazene (XIV). However, when a mixture of ammonia and formaldehyde was treated with a diazonium salt, the only product was observed to be the 3,7-bis(arylazo)-1,3,5,7-tetraazabicyclo[3,3,1]nonane (XV) (18).

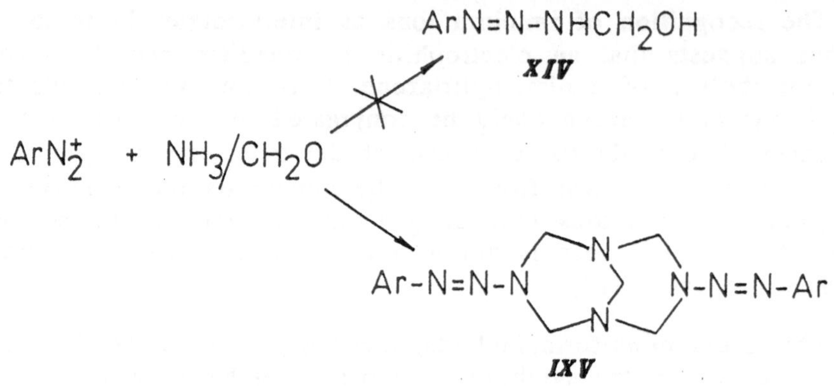

$$ArN=N-NHCH_2OH$$

XIV

$$ArN_2^+ \; + \; NH_3/CH_2O$$

$$Ar-N=N-N \underset{N}{\overset{N}{\diagdown}} N-N=N-Ar$$

IXV

ANTI-TUMOUR ACTIVITY

The hydroxymethyltriazenes display a broad spectrum of activity against several tumours, including the TLX5 lymphoma, the P388 leukemia and the ADJ/PC6 plasma cytoma (19) (Table 1). Anti-tumour activity amongst the HMT derivatives is variable. For example, the acetoxymethyltriazenes show the same spectrum of activity as the hydroxymethyltriazenes. On the other hand, the methoxymethyltriazenes (VIII) are active on the TLX5 but not on the P388 tumour, and show no in vitro cytotoxicity to cells growing in culture.

This behaviour becomes more understandable when the hydrolysis properties of the respective triazenes are examined (Table 2). The half-lives of the acetoxymethyltriazene and hydroxymethyl-triazene are virtually identical with that of the monomethyltriazene, whereas the methoxymethyltriazene does not break down at all in buffer just as a dimethyltriazene is resistant to hydrolysis.

The facility of hydrolysis of the acetoxymethyltriazene correlates with the reaction pathway in Scheme 4. Loss of acetate ions is fast followed by rapid hydrolysis of the iminium ion to give the hydroxymethyltriazene, which in turn loses formaldehyde rapidly to give the monomethyltriazene. The rate determining step in the whole process is the slow hydrolysis of the monomethyltriazene to the arylamine, nitrogen and methanol. Thus the monomethyl-triazene, the hydroxymethyltriazene and the acetoxymethyltriazene afford exactly the same kinetic profile when the decay is followed by UV spectroscopy. In contrast, the methoxymethyltriazene does not decompose at neutral pH because of the poor leaving group character of the methoxyl group.

TABLE 1

ANTI-TUMOUR ACTIVITY

X—⟨ ⟩—$N=N-N\begin{smallmatrix}R\\Me\end{smallmatrix}$

X	R	Tumour	% ILS (T/C)	Optimal dose (mg/Kg)
MeO_2C	H	TLX5[1]	87	5
MeO_2C	Me	TLX5	53	80
MeO_2C	CH_2OH	TLX5	120	5
MeO_2C	CH_2OBz	TLX5	31	20
MeO_2C	CH_2OAc	TLX5	56	40
		P388	67	120
Ac	CH_2OAc	P388	67	240
NO_2	CH_2OAc	P388	45	240
MeO_2C	CH_2OMe	TLX5	68	40
		P388	inactive	
Ac	CH_2OMe	P388	inactive	

1. 2×10^5 TLX5 lymphoma cells S.C. day 0 into CBA mice in groups of 5. Drugs administered days 3 to 7.

TABLE 2

CHEMICAL STABILITY

Half-life determinations of hydroxy-methyltriazene derivatives in phosphate buffer (pH 7.5) at 37°C by UV spectroscopy.

MeO_2C—⟨ ⟩—$N=N-N\begin{smallmatrix}R\\Me\end{smallmatrix}$

R	λ_{max}(nm)	$t_{1/2}$ (min)
H	313	12.1
Me	328	⟩24 h
CH_2OH	310	12.0
CH_2OAc	310	12.9
CH_2OMe	292	⟩24 h

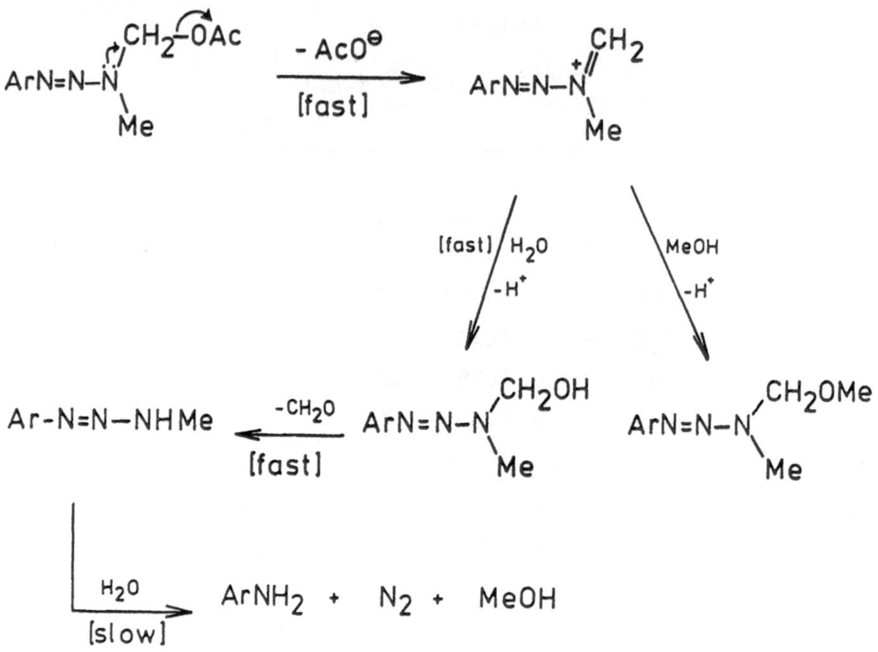

Scheme IV

In searching for a suitable pro-drug form of the hydroxy-methyltriazene, it seems that the acetoxymethyltriazene represents one extreme with a very labile leaving group in contrast to the very poor leaving group of the methyl ether. An in-between situation would be ideal and so we decided to look at S-alkyl, S-aryl and O-aryl ethers as potential pro-drugs. Acetoxymethyltriazenes resist solvolysis by an alkane thiol (Scheme 3), but Iley (20) has succeeded in preparing the alkylthiomethyltriazene (ArN=N-NMe-CH$_2$SR) by reaction of the hydroxymethyltriazene with a thiol/alcohol mixture in the presence of HCl. However, preliminary hydrolysis studies with these alkyl sulfides shows that they do not break down in buffer.

Reaction of the acetoxymethyltriazene with the mercaptan is facilitated by conversion to the sulfur anion. Thus the acetate reacts readily with the sodium thiophenolate in dimethylformamide to give a series of 1-aryl-3-arylthiomethyl-3-methyltriazenes (XVI) (21):

The oxygen analogues, XVII, can be obtained in similar reaction of the acetoxymethyltriazene with the sodium phenolate in dimethyl formamide, but the formation of the ether is accompanied by a second triazene product identified as the phenylmethyltriazene, XVIII. These isomeric products are also formed together when the acetate is reacted with the phenol itself in chloroform (22).

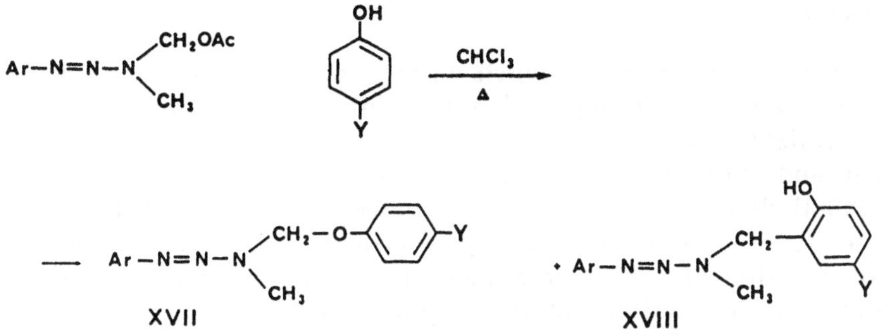

Unlike the O-alkyl ethers (VIII), which are stable in buffer, the O-aryl ethers (XVII) undergo hydrolysis in pH 7.5 buffer. Furthermore, the stability of the O-aryl ether is exquisitely sensitive to the electronic character of the substituent in the O-aryl moiety. Thus the p-nitrophenoxymethyltriazene (XVIIa) hydrolyses very rapidly with a half-life of ca. 1 min. to the hydroxymethyltriazene. The rate of the initial reaction is measured by following the rate of growth of absorption at 400 nm, the absorption maximum of the p-nitro-phenolate ion. This is the first time to our knowledge that the rate of reaction has been measured for the actual decomposition of a hydroxymethyltriazene derivative. In contrast, the p-bromo-phenoxymethyl-triazene (XVIIb) decomposes slowly with a half-live of approx. 60 min. (Scheme 5).

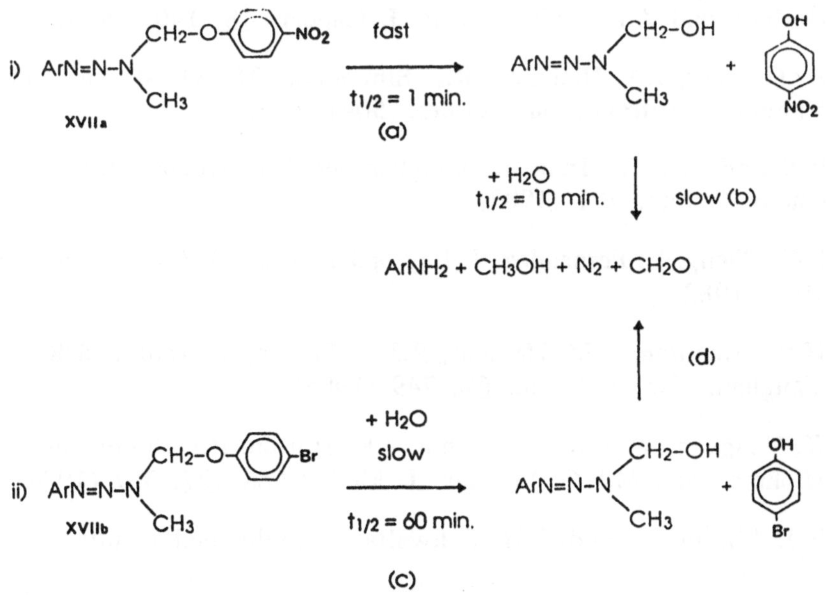

Scheme V

11

CONCLUSION

Evidently, hydroxymethyltriazenes are not as elusive chemical species as was once thought. Stable hydroxymethyltriazenes are readily available and a variety of derivatives with a full range of chemical and biological stability have been synthesised. The hydroxymethyltriazenes, and the derivatives that we have studied, have anti-tumour activity against experimental mouse tumours; the level of activity of the hydroxymethyltriazene derivatives is at least comparable to the activity of dimethyltriazenes such as DTIC and the 1-p-carboxyphenyl-3,3-dimethyltriazene presently undergoing clinical trial. Some of these active derivatives have the stability in aqueous buffer to warrant further investigation as pro-drugs, and certainly these compounds should be considered as clinical alternatives to DTIC and mitozolomide

REFERENCES

1. K. Vaughan and M.F.G. Stevens. Chem. Soc. Review. 7, 377 (1978)

2. T.P. Ahern and K. Vaughan. J. Chem. Soc. Chem. Comm. 701 (1973).

3. T.P. Ahern, H. Fong and K. Vaughan. Can. J. Chem. 55 1701 (1977).

4. K. Vaughan. J. Chem. Soc. Perkin Trans. II. 17 (1977).

5. C.J. Rutty, D.R. Newell, R.B. Vincent, G. Abel, P.M. Goddard, S.J. Harland and A.H. Calvert. Brit. J. Cancer. 48, 140 (1983).

6. A. Gescher, J.A. Hickman, R.J. Simmonds, M.F.G. Stevens and K. Vaughan. Tetrahedron Letters. 5041 (1978).

7. R.J. LaFrance, Y. Tang, K. Vaughan and D.L. Hooper. J. Chem. Soc. Chem. Comm. 721 (1983).

8. S.C. Cheng, L. Fernandez, J. Iley and E. Rosa. J. Chem. Res. (S). 108 (1983).

9. H.W. Manning, C.M. Hemens, R.J. LaFrance, Y. Tang and K. Vaughan. Can. J. Chem. 62, 749 (1984).

10. K. Vaughan, Y. Tang, G. Llanos, J.K. Horton, R.J. Simmonds, J.A. Hickman and M.F.G. Stevens. J. Med. Chem. 27, 357 (1984).

11. R.J. Simmonds and C.H. Schwalbe. Unpublished results.

12. N.W. Gibson, J. Hartley, R.J. LaFrance and K. Vaughan. Carcinogenesis. 7, 259 (1986).

13. C.M. Hemens, H.W. Manning, K. Vaughan, R.J. LaFrance and Y. Tang. Can. J. Chem. 62, 741 (1984).

14. C.M. Hemens and K. Vaughan. J. Chem. Soc. Perkin Trans II. 11 (1986)

15. K. Vaughan, K.U.K. Gamage Nicholas, R.D. Singer, M. Roy and N.W. Gibson. Anti-Cancer Drug Design. 2, 279 (1987).

16. M. Overton, J.A. Hickman, M.D. Threadgill, K. Vaughan and A. Gescher. Biochem. Pharmacol. 34, 2055 (1985).

17. R.J. LaFrance, H.W. Manning and K. Vaughan. J. Org. Chem. 50, 2229 (1985).

18. R.D. Singer, K. Vaughan and D.L. Hooper. Can. J. Chem. 64, 1567 (1986).

19. L.M. Cameron, R.J. LaFrance, C.M. Hemens, K. Vaughan, R. Rajaraman, D.C. Chubb and P.M. Goddard. Anti-Cancer Drug Design. 1, 27 (1985).

20. J. Iley, E. Rosa and L. Fernandes. J. Chem. Res. (S). 264 (1987).

21. K. Vaughan, H.W. Manning, M.P. Merrin and D.L. Hooper. Can. J. Chem. 66, 2487 (1988).

22. M.P. Merrin, R.J. LaFrance and K. Vaughan. Unpublished results.

MECHANISMS OF THE BIOLOGICAL ACTIONS OF TRIAZENES

M. J. TISDALE

Cancer Research Campaign Experimental Chemotherapy
Group, Pharmaceutical Sciences Institute, Aston
University, Birmingham B4 7ET, UK

5-(3,3-Dimethyl-1-triazeno)-imidazole-4-carboxamide (DTIC, Fig. 1) was the first member of the series of antitumour triazenes and was chosen for clinical trial because of its activity against murine leukaemia L1210. Significant activity was demonstrated against malignant melanoma with an overall response rate of 21%, while marginal activity was demonstrated against other solid tumours.

Whilst DTIC was originally considered as an inhibitor of the purine de novo biosynthesis pathway, mechanistic studies soon revealed that antitumour activity was due to the presence of the triazene moiety. Although DTIC has been shown to be more cytotoxic in the presence of light, due to decomposition to dimethylamine and 5-diazoimidazole-4-carboxamide (2), this is considered to be a non-specific toxification mechanism not related to the mechanism of cytotoxicity. In the absence of light an alternative decomposition pathway is followed with the initial loss of a methyl group to form 5-(3-methyl-1-triazeno)imidazole-4-carboxamide (MTIC) (Fig. 1), which was considered to be the activated form of the molecule able to undergo further decomposition in cancer cells with the ultimate formation of 5-aminoimidazole-4-carboxamide (AIC) and a methyl carbonium ion, which then interacts with DNA (3) (Fig. 1). Bioactivation of DTIC occurs by enzymatic hydroxylation of one of the methyl groups yielding initially 5-(3-hydroxymethyl-3-methyl-1-triazeno)imidazole-4-carboxamide (HMIC), which has been postulated to act as a transport form of the DTIC-derived methylating agent (4). Alkylation of DNA results in the formation mainly of N^7-methylguanine, with smaller amounts of O^6-methylguanine (5). Formation of the latter base is probably more important in the cytotoxicity of the antitumour triazenes.

The recognition that the antitumour activity of DTIC was a property only of its triazene function led to the synthesis of a wide range of aryl dimethyltriazenes which were active against a broad spectrum of animal tumours displaying activity against tumours naturally insensitive to either alkylating agents or antimetabolites (6). In this property they closely resemble the nitrosoureas, agents also known to cause alkylation of DNA with the formation of O^6-alkylated guanine (7). Antitumour activity is only observed when the leaving group in the triazene moiety is CH_2CH_2Cl or CH_3, while C_2H_5 and higher homologues are devoid of antitumour activity (8). In this respect, also, the antitumour triazenes closely resemble the nitrosoureas.

Triazenes, Edited by T. Giraldi *et al.*
Plenum Press, New York, 1990

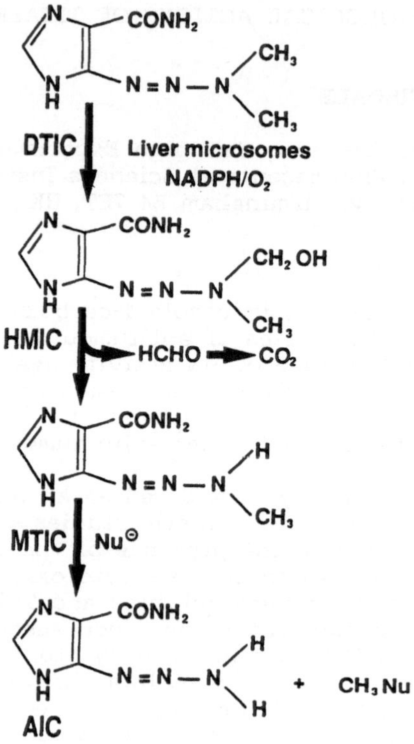

Figure 1 - Metabolic activation and alkylation of nucleophilic centres by DTIC.

While controversy surrounds the hypothesis that monomethyltriazenes are the active metabolites of the antitumour dimethyltriazenes (9,10), a generally accepted hypothesis for the mechanism of action of dimethyl-triazenes as antitumour agents is that they alkylate DNA (11). However, it is difficult to explain the absolute requirement for a methyl group at the terminal nitrogen of the triazene moiety in order to elicit antitumour activity, since monoethyltriazenes are also alkylating agents forming O^6-ethylguanine and their diethyltriazene precursors are dethylated by liver homogenates (6). Although ethyltriazenes form O^6-ethylguanine, the amount formed is quantitatively less than with methyltriazenes and cells differing in their capacity to repair O^6-alkylguanine lesions do show differences in the response to the cytotoxic effects of methyl and ethyltriazenes (12). Thus cells deficient in their repair capacity (Mer$^-$) show an increased cytotoxic response to both a monomethyl and a monochloroethyl triazene when compared with a repair proficient cell line (Mer$^+$), while no difference in toxicity between Mer$^-$ and Mer$^+$ cell lines was observed with an ethyl-triazene. Also in a series of cell lines the ability to repair O^6-alkylguanine is directly related to resistance to MTIC (13). This strongly suggests the importance of the O^6-methylguanine lesion in the mechanism of cytotoxicity of the methyltriazenes. The subsequent reactions responsible forcell death are not known but neither the formation of DNA single-strand breaks or DNA protein cross-links can account for the differential cytotoxicity observed in Mer$^+$ and Mer$^-$ cells towards the monomethyltriazenes (12). In contrast DNA interstrand cross-linking appears to be implicated in the mechanism of cytotoxicity of the chloroethyltriazenes. The initial reaction of the chloroethyl group is with the O^6-position of guanine moieties in DNA to give O^6-2-chloroethylguanine followed by migration of the alkyl group to the N^1-position to form a reactive intermediate O^6-ethanodeoxyguanosine (II) followed by formation of a cross-link to the N^3-position of a cytosine in the opposite DNA strand (Fig. 2).

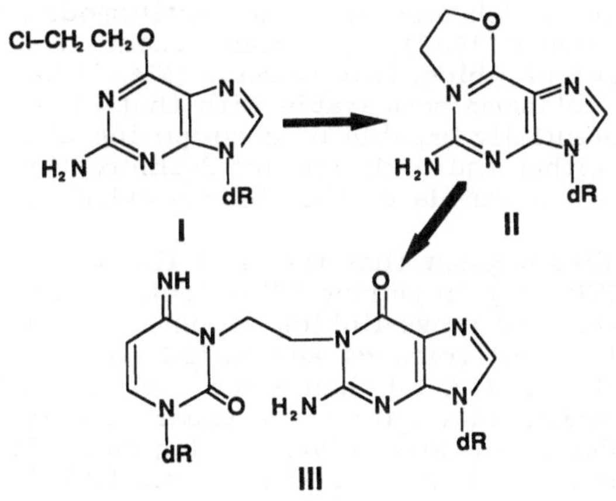

Figure 2   Formation of a cCydCh$_2$Ch$_2$dGuo cross-link in DNA.

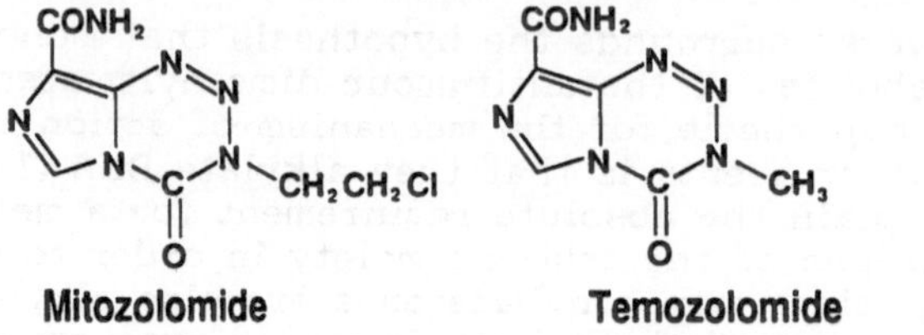

Figure 3   Structures of imidazotetrazinones.

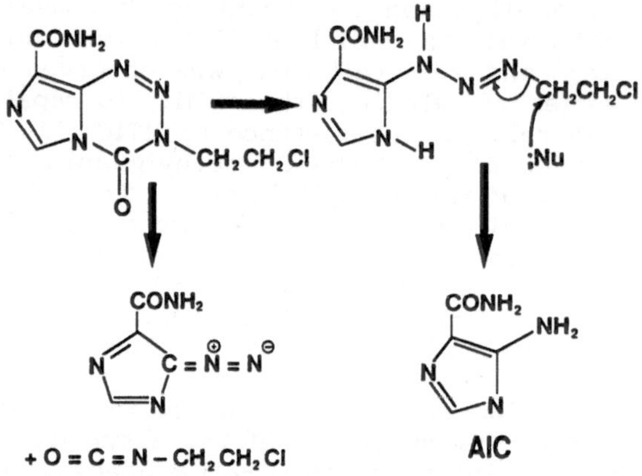

Figure 4   Potential decomposition pathways of mitozolomide.

We have recently identified a series of imidazo [5,1-d]-1,2,3,5-tetrazine derivatives which display potent antitumour activity against a range of murine tumours (14,15) and human tumour xenografts (16), the chloroethyl analogue of which, Mitozolomide (Fig. 3) has an activity against malignant melanoma comparable with that of DTIC (17). Mitozolomide is potentially capable of decomposing along two pathways (Fig. 4) producing either the toxic species 2-chloroethylisocyanate and 5-diazoimidazole-4-carboxamide or the chloroethyltriazene (MCTIC).

In vitro studies suggest that the most likely antitumour metabonate is the triazene MCTIC (18). In murine L1210 leukaemia cells mitozolomide produces DNA interstrand cross-linking which correlates with the cytotoxicity (19). Both mitozolomide and MCTIC are 5- to 6-fold more· toxic to a Mer[-] cell line (VA-13) than a Mer[+] cell line (IMR-90) and a concentration-dependent DNA interstrand cross-link formation is detected in VA-13 cells, which is linearly related to log cell kill, while in IMR-90 cells no interstrand cross-linking occurs (20). These data suggest that DNA interstrand cross-link formation may be a common mechanism for the in vitro cytotoxicity of both mitozolomide and MCTIC. Pretreatment of four Mer$^+$ human cells with the DNA methylating agent streptozotocin causes a dramatic increase in the sensitivity to the cytotoxic effects of mitozolomide (21). Streptozotocin apparently saturates the monoadduct repair system and allows mitozolomide to form interstrand cross-links in these cells. These results strengthen the hypothesis that mitozolomide induced cross-links probably arise after an initial akylation at the O^6-position of guanine residues in DNA.

Alkylation of guanine residues in DNA at the O^6-position is also important in the cytotoxic action of the methylimidazotetrazinone, Temozolomide (Fig.3), which is capable of cleavage to form a linear triazene, MTIC, the putative active metabolite generated by host metabolism of DTIC. Thus cell lines with constitutive levels of O^6-methylguanine-DNA methyltransferase (O^{6}MeGMt) are less sensitive to the cytotoxic effects of temozolomide than cells lacking the repair enzyme. In addition Mer$^+$ cells depleted of their O^{6}MeGMT by incubation with the free base O^6-methylguanine show an increased sensitivity to temozolomide, but not to the ethyl analogue (Fig. 3), which also displays no differential toxicity between Mer$^+$ and Mer$^-$ cell lines (22). These results suggest that ethyltriazenes probably produce a non-specific cytotoxicity by a different cytotoxic lesion from the methyl analogues.

The methyl and ethyl imidazotetrazinones also differ in their ability to induce differentiation, as measured by haemoglobin synthesis in K562 human erythroleukaemia cells (23,24). Thus, while both temozolomide and a monomethyltriazene are capable of inducing an increase in the number of benzidine positive cells, the corresponding ethyl analogues have no effect, even at concentrations causing an equivalent inhibition of cell growth. Induction of erythroid characteristics is accompanied by a decrease in the total 5-methylcytosine content of DNA, suggesting that differentiation is due to gene hypomethylation (25). However, this hypothesis has recently been questioned (24) based on the methylation status of the ε globin gene, globin gene, c-myc and c-ras oncogenes using the methylation sensitive restriction endonucleases, MspI, HpaII and HhaI. However, care should be taken in the measurement of the methylation status of a particular gene using restriction endonucleases since these enzymes will not cleave nucleotides containing O^6-methylguanine in the restriction site (26).

Temozolomide has also been shown to cause a decrease in the level of 5-methylcytosine in cell lines without an accompanying morphological or functional differentiation, and the effect occurs at lower drug concentrations in a chemosenstive (Mer$^-$) cell line (27). In order to investigate the mechanism of hypomethylation of DNA the possibility of inhibition of DNA methylase by either the alkylated DNA or by the free drug has been considered. While temozolomide alone has no effect on DNA methylase activity, in vitro studies show alkylated DNA to inhibit the transfer of methyl groups from S-adenosyl-L-methionine to M.lysodeikticus DNA by purified DNA methylase (27). In addition DNA isolated from a Mer$^-$ cell line previously treated with temozolomide also inhibits the methylase reaction (28). The maximum effect is observed 6h after drug addition and is proportional to the concentration of temozolomide to which the cells had previously been exposed.

The potential role, if any, of the inhibition of DNA methyltransferase to the mechanism of cytotoxicity of the methyltriazenes and imidazotetrazinones must remain speculative. DNA methyltransferase is a S-phase specific enzyme (29) and in analogy with other S-phase specific enzymes, the inhibition might be expected to block cells in the late S or G_2 phase of the cell cycle. Indeed flow cytometric studies of L1210 leukaemia cells treated with temozolomide do show an arrest in the late S/G_2 phase of the cell cycle (30). Inhibition of DNA methylation by 5-aza-2'-deoxycytidine has been suggested as a possible mechanism of chemotherapeutic action (31), although this agent can also produce its lethal effects by incorporation into DNA (32). The block to enzyme activity produced by temozolomide treated DNA appears to be reversible and cells escaping the block would have hypomethylated DNA, and the potential of displaying aberrant gene expression and malignant progression. However, further studies are required to fully elucidate the mechanism of events leading to cell death by this novel group of antitumour agents.

Acknowledgement - This work has been supported by a grant from the Cancer Research Campaign.

REFERENCES

1. S.K. Carter and M.A. Friedman, 5-(3,3-dimethyl-1-triazeno)-imidazole-4-carboxamide (DTIC, DIC, NSC-45388) A new antitumour agent with activity against malignant melanoma, Eur. J. Cancer 8:85 (1972).

2. A.H. Girculath and T.L. Loo, Mechanism of action of 5-(3,3-dimethyl-1-triazeno)-imidazole-4-carboxamide in mammalian cells in culture, Biochem. Pharmacol. 21:2335 (1972).

3. H.T. Nagasawa, F.N. Shirota and N.S. Mizuno, The mechanism of alkylation of DNA by 5-(3-methyl-1-triazeno)imidazole-4-carboxamide (MIC), a metabolite of DIC (NSC-45388). Non-involvement of diazomethane, Chem.-Biol. Interactions, 8: 403 (1974).

4. G.F.Kolar, M. Maurer and M. Wildschütte, 5-(3-Hydroxymethyl-3-methyl-1-triazeno)imidazole-4-carboxamide (DIC, DTIC, NSC-45388) Cancer Lett. 10:235 (1980).

5. L. Meer, R.C. Janzer, P. Kleihues and G.F. Kolar, In vivo metabolism and reaction with DNA of the cytostatic agent, 5-(3,3-dimethyl-1-triazeno)imidazole-4-carboxamide (DTIC), Biochem. Pharmacol. 35:3243 (1986).

6. R.C.S. Audette, T.A. Connors, H.G. Mandel, K. Merai and W.C.J. Ross, Studies on the mechanism of action of the tumour inhibitory triazenes, Biochem. Pharmacol. 22:1855 (1973).

7. P.D. Lawley, DNA as a target of alkylating carcinogens, Brit. Med. Bull. 36:1 (1980).

8. T.A. Connors, P.M. Goddard, K. Merai, W.C.J. Ross and D.E.V. Wilman, Tumour inhibitory triazenes: structural requirements for an active metabolite, Biochem. Pharmacol. 25:241 (1976).

9.   A. Gescher, J.A. Hickman, R.J. Simmonds, M.F.G. Stevens and K. Vaughan, Studies on the mode of action of antitumour triazenes and triazines - II. Investigation of the selective toxicity of 1-aryl-3,3-dimethyltriazenes, Biochem. Pharmacol. 30:89 (1981).

10. G. Sava, S. Zorzet, L. Perissin, T. Giraldi and L. Lassiani, Effects of an inducer and an inhibitor of hepatic metabolism on the antitumour action of dimethyltriazenes, Cancer Chemother. Pharmacol. 21:241 (1988).

11. V.H. Bono, Jr., Studies on the mechanism of action of DTIC (NSC-45388), Cancer Treat. Rep. 60:141 (1976).

12. N.W. Gibson, J. Hartley, R.J. LaFrance and K. Vaughan, Diferential cytotoxicity and DNA-damaging effects produced in human cells of the Mer$^+$ and Mer$^-$ phenotypes by a series of alkyltriazenylimidazoles, Carcinogenesis 7:259 (1980).

13. J.M. Lunn and A.L. Harris, Cytotoxicity of 5-(3-methyl-1-triazeno)-imidazole-4-carboxamide (MTIC on Mer$^+$, Mer$^+$Rem$^-$ and Mer$^-$ cell lines: Differential potentiation by 3-acetamidobenzamide, Br. J. Cancer 57:54 (1988).

14. J.A. Hickman, M.F.G. Stevens, N.W. Gibson, S.P. Langdon, C. Fizames, F. Lavelle, G. Atassi, E. Lunt and R.M. Tilson, Experimental antitumour activity against murine tumour model systems of 8-carbamoyl-3-(2-chloroethyl)imidazo[5,1-d]-1,2,3,5-tetrazin-4(3H)-one (Mitozolomide), a novel broad-spectrum agent. Cancer Res., 45:3008 (1985).

15. M.F.G. Stevens, J.A. Hickman, S.P. Langdon, D. Chubb, L. Vickers, R.Stone, G. Baig, C. Goddard, N.W. Gibson, J.A. Slack, C. Newton, E. Lunt, C. Fizames and F. Lavelle, Antitumor activity and pharmacokinetics in mice of 8-carbamoyl-3-methyl-imidazo[5,1-d]-1,2,3,5-tetrazin-4(3H)-one (CCRG 81045; M and B 39831), a novel drug with potential as an alternative to dacarbazine, Cancer Res. 47: 5846 (1987).

16. O. Fodstad, S. Aamdal, A. Pihl and M.R. Boyd, Activity of Mitozolomide (NSC 353451) a new imidazotetrazinone, against xenografts from human melanomas, sarcomas and lung and colon carcinomas, Cancer Res. 45:1778 (1985).

17. S. Gundersen, S. Aamdal and O. Fodstad, Mitozolomide (NSC 353451), a new active drug in the treatment of malignant melanoma. Phase II trial in patients with advanced disease, Br. J. Cancer 55:433 (1987).

18. C.M.T. Horgan and M.J. Tisdale, Antitumour Imidazotetrazinones -IV. An investigation into the mechanism of antitumour activity of a novel and potent antitumour agent, mitozolomide (CCRG81010, M and B 39565, NSC 353451), Biochem. Pharmacol. 33:2185 (1984).

19. N.W. Gibson, L.C. Erickson and J.A. Hickman, Effects of the antitumour agent 8-carbamoyl-3-(2-chloroethyl)imidazo[5,1-d]-1,2,3,5-tetrazin-4(3H)-one on the DNA of mouse L1210 cells. Cancer Res. 44:1767 (1984).

20. N.W. Gibson, J.A. Hickman and L.C. Erickson, DNA cross-linking and cytotoxicity in normal and transformed human cells treated in vitro with 8-carbamoyl-3-(2-chloroethyl)imidazo[5,1-d]-1,2,3,5-tetrazin-4(3H)-one, Cancer Res. 44:1772 (1984).

21. N.W. Gibson, J.A Hartley, D. Barnes and L.C. Erickson, Combined effects of streptozotocin and mitozolomide against four human cell lines of the Mer[+] phenotype, Cancer Res. 46:4995 (1986).

22. M.J. Tisdale, Antitumour imidazotetrazines - XV. Role of guanine O^6 alkylation in the mechanism of cytotoxicity of imidazotetrazinones, Biochem. Pharmacol. 36:457 (1987).

23. M.J. Tisdale, Induction of haemoglobin synthesis in the human leukaemia cell line K562 by monomethyltriazenes and imidazotetrazinones, Biochem. Pharmacol. 34: 2077 (1985).

24. M. Zucchetti, C.V. Catapano, S. Filippeschi, E. Erba and M. D'Incalci, Temozolomide induced differentiation of K562 leukaemia cells is not mediated by gene hypomethylation, Biochem. Pharmacol. 38:2069 (1989).

25. M.J. Tisdale, Antitumour Imidazotetrazinones - X. Effect of 8-carbamoyl-3-methylimidazo[5,1-d]-1,2,3,5-tetrazin-4-(3H)-one (CCRG 81045, M and B 39831, NSC 362856) on DNA methylation during induction of haemoglobin synthesis in human leukaemia cell line K562, Biochem. Pharmacol. 35:311 (1986).

26. R.S. Wu, S. Hurst-Calderone and K.W. Kohn, Measurement of O^6-alkyl-guanine-DNA alkyltransferase activity in human cells and tumor tissues by restriction endonuclease inhibition, Cancer Res. 47:6229 (1987).

27. M.J. Tisdale, Antitumour imidazotetrazinones and gene expression, Acta Oncologica 27:511 (1988).

28. M.J. Tisdale, Antitumour imidazotetrazinones - XVIII. Modification of the level of 5-methylcytosine in DNA by 3-substituted imidazo-tetrazinones, Biochem. Pharmacol. 38:1097 (1989).

29. M.C. Vogel, T. Papadopoulos, H.K. Müller-Hermelink, D. Drahovsky and G.P. Pfeifer, Intracellular distribution of DNA methyltransferase during the cell cycle, FEBS Lett. 236:9 (1988).

30. C.V. Catapano, M. Broggini, E. Erba, M. Ponti, L. Mariani, L. Citti and M. D'Incalci, In vitro and in vivo methazolastone-induced DNA damage and repair in L1210 leukaemia sensitive and resistant to chloroethylnitrosoureas, Cancer Res. 47:4884 (1987).

31. V.L. Wilson, P.A. Jones and R.L. Momparier, Inhibition of DNA methylation in L1210 leukemic cells by 5-aza-2'-deoxycytidine as a possible mechanism of chemotherapeutic action, Cancer Res. 43:3493 (1983).

32. J. Vesley and A. Cihak, Incorporation of a potent antileukemic agent, 5-aza-2'-deoxycytidine into DNA of cells from leukemic mice, Cancer Res. 37:3684 (1977).

TRIAZENES AND TRIAZENE N-OXIDES:

ANTITUMOUR ACTION IN ANIMAL TUMOUR SYSTEMS

Derry E. V. Wilman

Drug Development Section
Institute of Cancer Research
Cancer Research Campaign Laboratory
Cotswold Road, Sutton
Surrey SM2 5NG, U.K.

INTRODUCTION

The wealth of literature relating to the triazenes amounts, according to the Chemical Abstracts Service, to some 1200 papers over the past twenty years, of which many relate to the antitumour action of these compounds. Thus it is necessary to be selective in the choice of topics for a review of this nature. The object therefore is to attempt to show how the *in vivo* antitumour test results of these compounds have given an insight into their mode of action and the structural requirements for it, and led to the choice of clinical agents. Some significant observations have so far been largely ignored and these will be highlighted. Finally the activity of some triazene *N*-oxides will be discussed together with their potential application.

Although the first triazene synthesis was originally described by Griess in 1862 (1) and that of an aryldialkyltriazene in 1875 by Baeyer & Jaeger (2), their biological activity was not reported until much later. Most of the early investigations of the biological activity of the triazenes related to their carcinogenicity and was well reviewed by Druckrey (3). The first report of antitumour activity was made by Stock and his colleagues in 1955 (4). They showed that 3,3-dimethyl-1-phenyltriazene, 3,3-dimethyl-1-(4-nitrophenyl)triazene and 3,3-dimethyl-1-(4-tolyl)triazene inhibited the growth of the mouse sarcoma 180. The following year they went on to show that the same 1-aryl-3,3-dimethyltriazenes inhibited the mouse leukaemia 82 (5). Surprisingly, no further investigation of the aryldialkyltriazenes as antitumour agents was undertaken until fourteen years later. A 1969 South African patent (6) described over sixty triazenes (I) some of which had activity against tumours of the breast, uterus, ovary, prostate and bladder at dose levels in the range of 1-5 gm/kg.

Other types of biological activity have been reported for the 1-aryl-3,3-dialkyltriazenes, such as pre-emergent herbicidal activity (7) and phytotoxicity (8).

Triazenes, Edited by T. Giraldi *et al.*
Plenum Press, New York, 1990

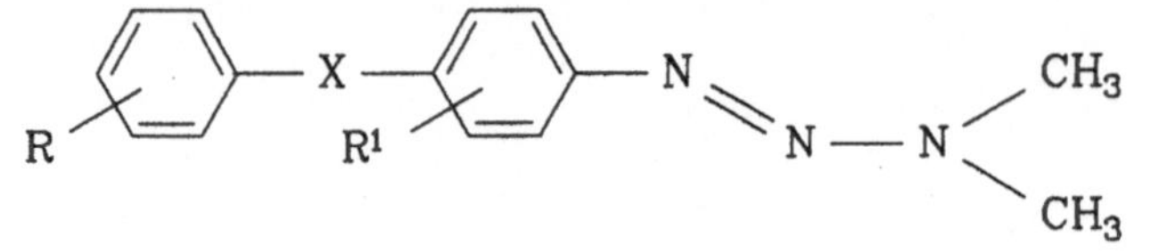

I

X is a direct C-C bond, CO, S, O, SO$_2$, N=N, CH$_2$, CH$_2$CH$_2$, CH=CH or C≡C

IMIDAZOLE TRIAZENES

The main thrust in this area, however, occurred in the early 1960s with the discovery of the antitumour activity of some heteroaryldialkyltriazenes. Shealy and his colleagues at the Southern Research Institute had shown that 5-diazoimidazole-4-carboxamide (III), formed by the diazotisation of 5-aminoimidazole-4-carboxamide (II, AIC) (Figure 1), was an inhibitor of the Ehrlich ascites tumour in mice and of the Walker 256 carcinosarcoma in rats (9). However, it was only slightly active against the sarcoma 180 and had no activity against the adenocarcinoma 755 or the L1210 leukaemia as shown in Table 1. Unfortunately, 5-diazoimidazole-4-carboxamide whilst stable in the solid state readily undergoes intramolecular cyclisation throughout the pH range in aqueous solution to give 2-azahypoxanthine (IV), which lacks antitumour activity. This same group went on to investigate ways of producing latent or prodrug forms of 5-diazoimidazole-4-carboxamide, amongst which were derivatives (V) obtained by coupling with aliphatic amines (11).

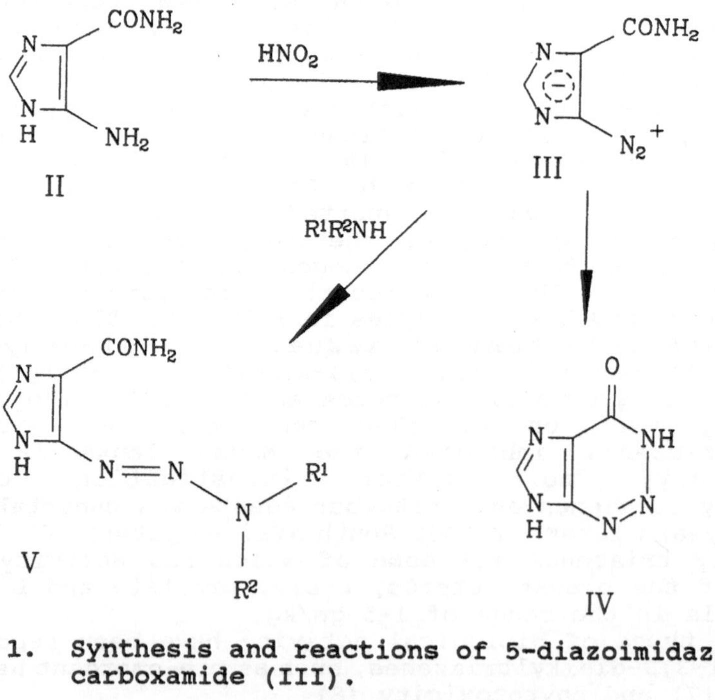

Figure 1. Synthesis and reactions of 5-diazoimidazole-4-carboxamide (III).

24

Table 1. Activity of 5-diazoimidazole-4-carboxamide in rodent tumour systems.

Dose (mg/kg daily, ip)	Schedule (days)	Tumour	Deaths	Average tumour weight (mg) or lifespan (days)		TWI[A] or ILS[B] (%)
				Treated	Controls	
15	1-11	Ehrlich	6/10	-	-	-
10		ascites	2/10	4.44	2.48	46
5		carcinoma	2/9	0.43	2.48	17
10	1-7	Sarcoma 180	3/5	-	-	-
5			2/6	555	996	55
2-5			1/6	532	996	53
1.3			0/6	609	996	61
2.5	1-11	Adenocarcinoma	5/10	-	-	-
1.3		755	2/10	659	1076	61
10	1-death	L1210	7/9	-	-	-
7.5		Leukaemia	0/10	8.7	9.3	0
5			0/10	9.7	9.3	4
20	1-11	Walker	5/6	-	-	-
10		adenocarcinoma	1/6	345	1646	21
5		256	0/6	1153	3032	38

[A] TWI = Tumour weight index
[B] ILS = Increase in life span
Adapted from reference 10

Two main classes of triazene were made initially, based on reaction with dimethylamine and *bis*(2-chloroethyl)amine to give 5-(3,3-dimethyl-1-triazeno)imidazole-4-carboxamide (VI, DTIC, DIC, Dacarbazine) (11) and 5-[3,3-*bis*(2-chloroethyl)-1-triazeno]imidazole-4-carboxamide (VII, BTIC, BIC) (12) respectively.

VI

DTIC was found to have substantial activity against the L1210 leukaemia, as shown in Table 2, producing up to 100% increase in life-span, when administered by various routes; intraperitoneal, subcutaneous, intramuscular or oral; and

Table 2. Activity of DTIC against the L1210 leukaemia.

Dose (mg/kg)	Route	Schedule (days)	Average Lifespan (days)		% ILS
			Treated	Control	
100	ip	1-30	17.7	9.0	96
100			15.0	8.9	68
120	ip	1-19	12.8	8.9	44
480	ip	1	15.5	8.0	93
500	ip	5	11.8	8.5	38
100	sc	1-30	14.2	9.0	57
100	im	1-30	14.5	9.0	61
100	iv	1-6	13.2	9.0	46
200	oral	1-30	15.7	8.2	91

Data from reference 10

schedules; single or multidose (10). This activity was also seen against a spectrum of other murine tumours, Table 3, although DTIC was inactive against the intracerebrally implanted L1210 leukaemia.

As a result of its broad spectrum of activity DTIC was chosen for clinical use, where it has proved to be the most active single agent against malignant melanoma (14). It also has the advantage that its final metabolite, AIC, is a natural product. It does, however, have the disadvantage of causing severe nausea and vomiting in all patients, which is poorly controllable by antiemetics.

The other significant product prepared from 5-diazo-imidazole-4-carboxamide, by Shealy's group, was 5-[3,3-*bis*(2-chloroethyl)-1-triazeno]imidazole-4-carboxamide (VII). This compound is even more potent than DTIC against many of the murine tumour systems. Indeed, the drug can be said to be curative of the L1210 leukaemia as animals treated with a single dose of 500 mg/kg survived for over eight months (12). Daily doses for up to 30 days led to major increases in life-span at doses in excess of 25 mg/kg. This activity is despite the extreme chemical instability of BTIC, due to spontaneous intramolecular cyclisation to a triazolinium chloride (VIII), as shown in Figure 2. The analogous 5-[3,3-*bis*(2-fluoroethyl)-1-triazeno]imidazole-4-carboxamide was also prepared and, although it cyclised considerably less readily, it was more toxic and much less active than BTIC (13). BTIC was also entered into clinical trial but proved to lack the antitumour activity in man which had been so evident in experimental animals.

Table 3. Activity of DTIC against other tumours.

Tumour	Dose (mg/kg) ip	Schedule (day)	Lifespan		ILS %
			Treated	Control	
L1210/MTX[A]	87	1-15	14.0	7.0	100
	50	1-death	12.5	7.0	78
L1210/MP[B]	125	1-15	11.0	7.0	57
L1210/TG[C]	75	1-15	18.5	10.0	85
P815	100	1-15	>20.8	14.3	>45[E]
P815/FU[D]	100	1-15	>28.6	14.7	>95[F]
L5178Y	75	1-15	12.2	9.2	32
					% of control
Sarcoma 180	100	1-7	results from 7 tests		23-50
Adenocarcinoma	125	1-11	results from 9 tests		3-44

[A] L1210 leukaemia resistant to methotrexate
[B] L1210 leukaemia resistant to 5-mercaptopurine
[C] L1210 leukaemia resistant to 6-thioguanine
[D] P815 leukaemia resistant to 5-fluorouracil
[E] Two out of ten 30-day survivors
[F] Five out of ten 30-day survivors
Data from references 10 and 13

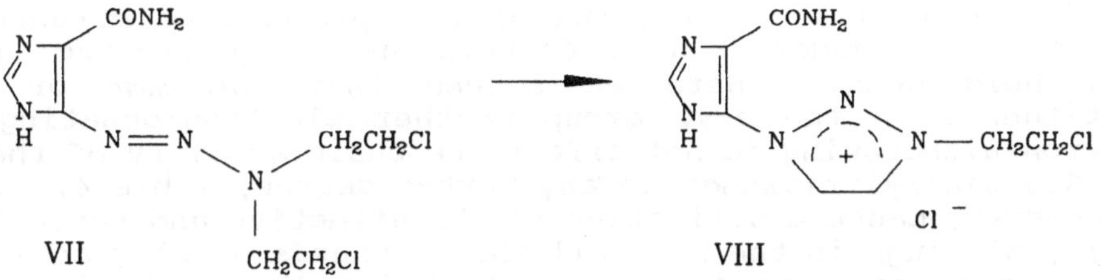

Figure 2. Intramolecular cyclisation of BTIC (VII).

The inter-relationship of the activities of DTIC and BTIC were studied by Kline and co-workers (15). A variant of the mouse leukaemia L1210 resistant to DTIC (L1210/45388/R) and a similar variant resistant to BTIC (L1210/82196/R) are described. The L1210/82196/R line was cross-resistant to DTIC and a number of structurally related compounds. The L1210/45388/R tumour was sensitive to BTIC and retained some sensitivity to the monomethyl analogue of DTIC, 5-(3-methyl-1-triazeno)imidazole-4-carboxamide, although it was resistant to other 3,3-disubstituted analogues of DTIC. Interestingly, the L1210/82196/R tumour line was also resistant to chloroethylnitrosoureas whereas the L1210/45388/R line was not.

In addition to DTIC and BTIC, Shealy's group investigated a wide range of triazenes derived from various heterocyclic amines, such as triazoles and pyrazoles and with considerable variation in the alkyl groups. Antitumour testing, of this wide range of heterocyclic triazenes against the L1210 leukaemia, indicated that only those with an N^3 methyl group were active (13). At the same time, this group also investigated a variety of 1-aryl-3,3-dialkyltriazenes which, although containing the same substituents as the imidazole derivatives and at least one methyl group, proved inactive or to have only modest activity against the same tumour (16, 17). This is in contrast to the subsequent observations of other groups.

ARYLTRIAZENES

Work in this area at the Institute of Cancer Research began in the early 1970s, under the direction of Connors and Ross (18, 19), with a two-fold aim. The ultimate aim was to produce a second-generation analogue of DTIC which would be more clinically acceptable, but initially it was necessary to establish the structural requirements of the triazenes for *in vivo* antitumour activity, as a basis for the design of such a drug. This investigation used the TLX5 lymphoma in mice as its test system and has been restricted to 1-aryl-3,3-dialkyltriazenes.

At that time the photolytic decomposition of DTIC to 5-diazoimidazole-4-carboxamide was thought responsible for the severe nausea and vomiting produced by this drug. As the 1-aryl-3,3-dialkyltriazenes are clinically stable and only decompose to a diazonium species under acid hydrolysis it was thought they might overcome this clinical problem. Regrettably, this has not proved to be the case indicating that the hydrolytic decomposition pathway alone is unlikely to be the cause of the toxicity.

An extensive series of these compounds has been produced over the years but the results of the qualitative structure-activity study undertaken by Connors and his colleagues (18) still hold good. Here we showed that *ortho*, *meta* or *para* substituents in the aryl group whether electron-donating or electron-withdrawing do not affect the basic activity of the 1-aryl-3,3-dialkyltriazenes to any marked degree, Table 4. They do, however, cause modification of the effective and toxic dose levels, although in the main all the compounds display activity over a range of 3 to 4 dose levels below the toxic dose, in antitumour tests against the mouse TLX5 lymphoma.

The data shown in Table 5 for the three 1-aryl-3,3-dimethyltriazenes is typical of that seen for most of these compounds. It shows that in going from an electron-donating substituent as in 1-(4-methoxyphenyl)-3,3-dimethyltriazene ($R = OCH_3$) to an electron-withdrawing substituent, 1-(4-methylsulphonylphenyl)-3,3-dimethyltriazene ($R = CH_3SO_2$), there is not a tremendous difference in activity. 1-(4-Carbamoylphenyl)-3,3-dimethyltriazene ($R = CONH_2$) lies between the others in terms of the electron distribution properties of the aryl group and has been the standard compound for our

Table 4. Antitumour activity of some 1-aryl-3,3-dimethyl-triazenes against the TLX5 lymphoma.

Substituent R	Max % ILS	Dose mg/kg (5 x daily)	
		Optimal	Toxic
H	53	64	128
o-COOH	68	32	64
m-COOH	62	100	200
p-COOH	72	25	200
o-COOCH$_3$	53	30	120
m-COOCH$_3$	63	100	400
p-COOCH$_3$	58	40	160
o-CONH$_2$	78	16	128
m-CONH$_2$	55	10	80
p-CONH$_2$	55	25	200
p-OCH$_3$	41	20	80
p-NO$_2$	39	100	400
p-CF$_3$	61	200	400
p-SO$_2$CH$_3$	80	80	320

Adapted from reference 18

investigations. However, there was some variation in the actual range of doses involved and the toxic dose. The toxic dose was usually from 2 to 8 times the optimal dose (the dose which produced the greatest increase in life span for the compound). These results were comparable with those seen for the heterocyclic triazenes reported by Shealy and his colleagues, but the activity was in general much greater than that group reported for the 1-aryl-3,3-dimethyltriazenes against the L1210 leukaemia (17).

The similarity in activity of this series of compounds discounted the possibility of the diazonion ion decomposition product having a part to play in their antitumour activity. The range of electron distribution effected by the different substituents, ranging from the electron-donating p-methoxy to the electron-withdrawing p-trifluoromethyl and p-methylsulphonyl derivatives, produce a similarly wide range of half-lives for the hydrolysis reaction (Table 5). This can vary from 11 min in the case of the p-methoxy derivative to greater than 90 days for the p-methylsulphonyl compound. In addition the diazonium ions proved to be more toxic than their analogous triazene. For instance, the diazonium salt derived from 1-(4-carbamoylphenyl)-3,3-dimethyltriazene is lethal at a single dose of 120 mg/kg but has no *in vivo* antitumour effect. Thus triazenes, which are

Table 5. Antitumour activity of some 1-aryl-3,3-dimethyl-triazenes towards the TLX5 lymphoma.

R	$t\frac{1}{2}$[A]	Dose (mg/kg) daily x 5	% ILS
OCH$_3$	11 mins	5	4
		10	6
		20	41
		40	39
		80	-35
CONH$_2$	86 days	12.5	14
		25	55
		50	51
		100	47
		200	4
		400	-61
SO$_2$CH$_3$	>90 days	10	2
		20	35
		40	74
		80	80
		160	43
		320	-43

[A] Half-life in 0.1 molar phosphate buffer at pH 7.4 and 37°C.

particularly hydrolytically unstable and will readily form significant amounts of the diazonium ion under physiological conditions, are unlikely to be useful antitumour agents as they may show increased toxicity with no gain in antitumour effect (18). This is evident in the data for 1-(4-methoxyphenyl)-3,3-dimethyltriazene (Table 5, R=OCH$_3$).

Of more interest were the results of varying the substituents at the N^3 position. The effects of doing this are shown in Table 6. The replacement of both methyl groups by higher alkyl groups leads to complete loss of activity, in contrast to the other forms of biological activity expressed by these compounds such as toxicity, carcinogenicity and mutagenicity which are all maintained, although specific sites may vary (3). Thus the presence of at least one methyl group at N^3 is a requirement for the expression of antitumour activity by these compounds. It also transpires that the remaining substituent at N^3 must be susceptible to ready metabolic or chemical loss to yield a monomethyltriazene, which is the putative active metabolite. Confirmation of the monomethyltriazene being the necessary metabolic product comes

Table 6. Antitumour activity of a series of 1-(4-carbamoyl-phenyl)-3,3-dialkyltriazenes against the TLX5 lymphoma.

Substituents		Max %	Dose (mg/kg) (daily x 5)	
R^1	R^2	ILS	Optimal	Toxic
CH_3	CH_3	55	25	200
C_2H_5	C_2H_5			400
$CH(CH_3)_2$	$CH(CH_3)_2$	Inactive		200
CH_3	CH_2CH_3	46	40	160
CH_3	CH_2CH_2OH	42	200	>400
CH_3	$CH_2CH_2CH_2CH_3$	78	50	200
CH_3	$(CH_2)_4CH_3$	102	40	160
CH_3	$CH_2C_6H_5$	86	160	320
CH_3	$C(CH_3)_3$	Inactive		>200
CH_3	H	43	30	60

Data from reference 18

in two ways. Firstly, 1-(4-carbamoylphenyl)-3-methyltriazene has *in vivo* antitumour activity in its own right whereas its 3-ethyl analogue is devoid of activity (Table 6). Secondly, 1-(4-carbamoylphenyl)-3-methyl-3-*tert*-butyltriazene is inactive although it contains an N^3-methyl substituent. The explanation for this is evident when viewed in metabolic terms. The only N^3 substituent in this molecule which has an α-proton, and is hence capable of undergoing the oxidative metabolism, which is the first step in N-alkyl metabolism, is the methyl group. Thus this is the group lost by metabolism to leave 1-(4-carbamoylphenyl)-3-*tert*-butyltriazene which, as the *tert*-butyl group is incapable of alkylation due to steric hindrance, has no worthwhile biological activity.

It was this latter result which confirmed our theory, based on our *in vivo* qualitative structure-active study, as to the minimal structural requirements for aryldialkyltriazenes to have antitumour activity. These are: a carrying structure at N^1 which may be aryl or heteroaryl; a methyl group at N^3; a group at N^3 which will be lost metabolically or chemically in preference to the methyl group (18). Although other theories of the mechanism of action, which do not require a monomethyltriazene intermediate, have been put forward no other has so far explained the inactivity of 1-(4-carbamoylphenyl)-3-methyl-3-*tert*-butyltriazene.

The large mass of data, arising from the qualitative structure-activity studies undertaken at Southern Research Institute, The Institute of Cancer Research and other places, predictably attracted the attention of Hansch and his colleagues. For separate series of triazenes based on an imidazole, pyrazole and benzene nucleus, they found equivalent relationships in each case for activity against the L1210 leukaemia. In each case log Po, the optimal partition coefficient for maximal antitumour activity, was found to be of the order of 1 (20). This result meant there was no room for improvement in potency by manipulation of the lipophilic character of the compounds. They also found that electron-donating substituents could be used to increase potency. However, as this also reduces the stability of the triazenes it is not a profitable manoeuvre. No special potency was associated with the heterocyclic rings. Thus there is no obvious way in which the potency of the 1-aryl-3,3-dialkyltriazenes may be increased. An alternative approach would be to reduce the toxicity of the compounds by structural manipulation. Regrettably, however, Hansch's group found this not to be a practical proposition in the case of the 1-aryl-3-alkyl-3-methyl-triazenes. The quantitative structure-activity relationships obtained for the toxicity of these compounds was so similar to that obtained for their antitumour activity that there is essentially no way in which more potent less toxic triazenes of this type may be synthesised (21).

Mention was made above that DTIC is inactive against the L1210 leukaemia if the tumour has been transplanted intracerebrally (13). Farquhar and Benvenuto (22) demonstrated that not all triazenes are inactive against the tumour implanted at this site, as shown in Table 7. As most had activity against the intraperitoneally implanted tumour but not against its intracerebral counterpart it is reasonable to assume that the blood-brain barrier (BBB) remained intact and that the tumour was also in the protected environment.

Two of the compounds investigated, 1-(4-carbamoylphenyl)-3,3-dimethyltriazene and 1-(4-carboxyphenyl)-3,3-dimethyltriazene, were active against the intracerebrally implanted tumour. The former was equally active in this situation as it was to the tumour implanted intraperitoneally. The properties of a drug for it to cross the blood-brain barrier have been established by Levin as high lipophilicity and a molecular weight below 400 (23). Whilst 1-(4-carbamoylphenyl)-3,3-dimethyltriazene meets these criteria, the 4-carboxy analogue does not when in its ionised form, and its activity therefore remains to be explained. However, 1-(4-carbamoylphenyl)-3,3-dimethyltriazene is also active against other intracerebrally implanted tumours and it is therefore reasonable to think that aryldimethyltriazenes, such as this, may be useful in the therapy of central nervous system metastases of malignant melanoma and of primary brain tumours.

Support for the ability of such triazenes to cross the blood-brain barrier and inhibit the growth of tumours in the brain was provided by an investigation in our own laboratory (24). A grade IV astrocytoma xenografted into the flank of immune deprived mice was sensitive to DTIC, aryltriazenes, nitrosoureas and other drugs used clinically to treat such tumours. The intracerebral implant of the tumour, however, was only sensitive to nitrosoureas and aryltriazenes lending further support for the application of aryltriazenes in the therapy of brain tumours, whether primary or metastatic.

Numerous attempts have been made to find carrying structures, onto which the dialkyltriazene moiety could be grafted, with the intention of the drug so formed being directed

Table 7. Comparison of the activity of some 1-aryl-3,3-dialkyltriazenes against the L1210 leukaemia implanted intraperitoneally or intracerebrally.

R	Max % ILS		Optimal daily dose (mg/kg)	
	ip	ic	ip	ic
o-CONH$_2$	10	10	65	40
m-CONH$_2$	50	22	39	40
p-CONH$_2$	40	45	39	160
o-COOCH$_3$	11	14	65	40
m-COOCH$_3$	30	10	39	160
p-COOCH$_3$	40	7	39	160
o-COOH	12	6	65	20
p-COOH	41	25	39	80
o-Cl	21	5	65	80
m-Cl	11	7	108	40
p-Cl	26	2	108	80
DTIC	50	18	108	160

IX

preferentially to the tumour. Examples of this approach are the use, as carrying structures, of acridines by Atwell and his colleagues (25), sulphonamides by Abel and his colleagues (26) and quinolines by Lin and Loo (27).

Several 3-(3,3-dialkyl-1-triazeno)acridines (IX) had excellent activity against the L1210 leukaemia (25). In this instance, however, a methyl group at N^3 was not a requirement for activity. For instance, the 3-methyl-3-propyl (IX; R^1 = CH$_3$, R^2 = C$_3$H$_7$) and 3,3-diethyl (IX; R^1 = R^2 = C$_2$H$_5$) analogues which are of comparable lipophilicity and toxicity also had equivalent antitumour activity. It is therefore likely that the usual routes of triazene metabolic activation are not of critical importance in these compounds. Further support for this contention comes from the associated observation that, the analogues in which the triazene moiety had been replaced by an azide group had even greater antitumour activity in this system.

The pH of tumours is reported to be lower than in normal tissue and that this difference may be enhanced by glucose administration (28). A compound of appropriate pKa could therefore be expected to be selectively deposited in the tumour, and indeed this occurs in the case of sulphapyrazine (29), and other sulphonamides including sulphadiazine (30). Connors and Ross have attempted to exploit these findings in the case of nitrogen mustard (31) and triazene derivatives (26) of a range of sulphonamides of varying pKa. In the latter case, although two of these triazenes, 2-[4-(3,3-dimethyl-1-triazeno)benzene-sulphonamido]pyrimidine (X, R = H) and 2-[4-(3,3-dimethyl-1-triazeno)benzenesulphonamido]-4-methylpyrimidine (X, R = CH$_3$) are amongst the best inhibitors of the TLX5 lymphoma so far observed, their activity did not appear to be related to any selective concentration in the tumour.

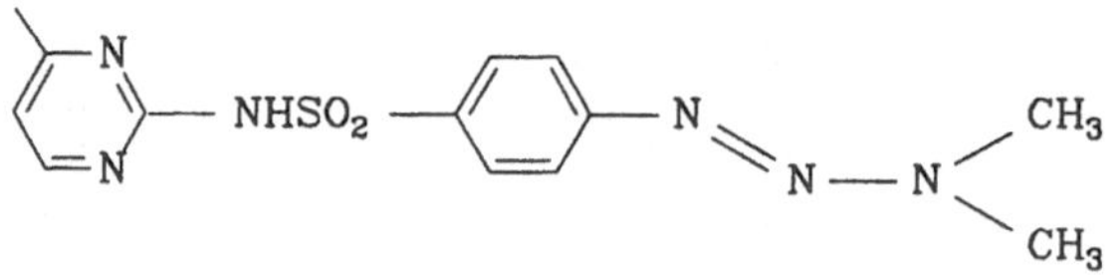

X

Sieber and Adamson (32) suggested that alkylating agents should be designed utilising carriers that are most likely to deliver the agents to specific target sites. These agents should also be designed with activity against specific types of tumour rather than as generalised cytotoxic agents. In this way it should be possible to increase the therapeutic index of these agents. Lin and Loo (27) attempted to exploit this idea by designing triazene derivatives which were specifically directed at melanine containing cells. Quinoline derivatives such as the chloroquins and chloropromazines show marked affinity for pigmented tissue and 4-(3-dimethylaminopropylamino)-7-[^{125}I]iodoquinoline concentrates in malignant melanoma (33). They therefore prepared a number of 6-, 7- and 8-halo substituted 4-(3,3-dimethyl-1-triazeno)quinolines. 8-Chloro-4-(3,3-dimethyl-1-triazeno)quinoline (XI) was the most active of these compounds against both P388 and L1210 tumours (27). In view of this activity the three 8-halo-4-(3,3-dimethyl-1-triazeno)quinolines together with DTIC and BTIC were tested against the B16 melanoma. Only BTIC showed significant activity.

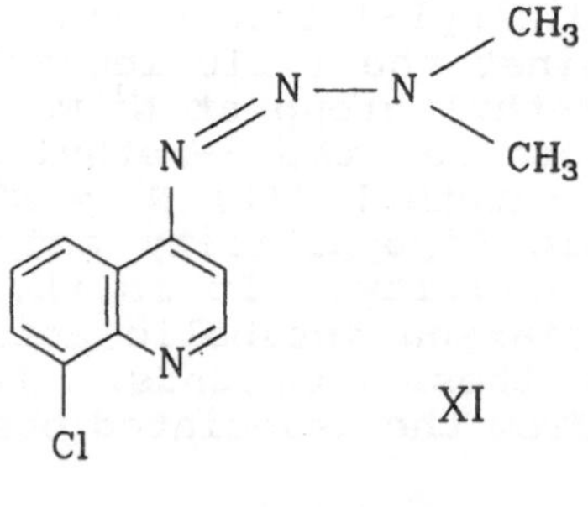

XI

Lin and Loo (27) found no correlation between the antitumour activity of these compounds and their ability to bind to synthetic melanin, but as the B16 is the only melanotic tumour amongst them the lack of antitumour activity in this system would predict such a result. However, these authors do not report on the binding of BTIC to synthetic melanin.

Another interesting feature associated with DTIC is its ability to cause antigenic changes in tumours and to reduce their oncogenic potential. Bonmasser and his colleagues (34) demonstrated that, after four different L1210 lines had been treated with DTIC for several generations a marked increase in the median survival times of untreated tumour bearing mice was observed. Schmid and Hutchinson (35) showed similar results with 1-phenyl-3,3-dimethyltriazene and 1-phenyl-3-methyltriazene and also that the tumours could be transplanted into hamsters.

However, perhaps the most interesting result of this type, applicable to chemotherapy, was the effect of *in vivo* treatment with DTIC on the L5178Y lymphoma. In this case after treatment with DTIC for four transplant generations the tumour was curable by a single dose of 1,3-*bis*(2-chloroethyl)-1-nitrosourea (BCNU), whereas the parent tumour was unaffected by BCNU (36). Similar results were also reported for the EL4 and Gross-virus induced leukaemias (37).

The mechanism of action of the aryldialkyltriazenes involves the intermediacy of a 1-aryl-3-hydroxymethyl-3-methyltriazene (XII) and a 3-aryl-1-methyltriazene (XIII) (Figure 3). Obviously either of these two metabolites would provide alternative drugs not requiring biological activation *in vivo*. The possible use of a stable monomethyltriazene has been explored (26), but in most cases the monomethyltriazene is not as active *in vivo* as its dimethyl parent (38). The *in vivo* activity is, however, linked to the chemical stability of the monomethyltriazene (18, 39). The *in vivo* stability is such that its decomposition to the monomethyltriazene is rapid and the rate-determining step of the hydrolysis is the hydrolysis of the monomethyltriazene (40).

Alternative prodrug forms of the hydroxymethyltriazenes have been investigated by Vaughan's group. These have included 1-

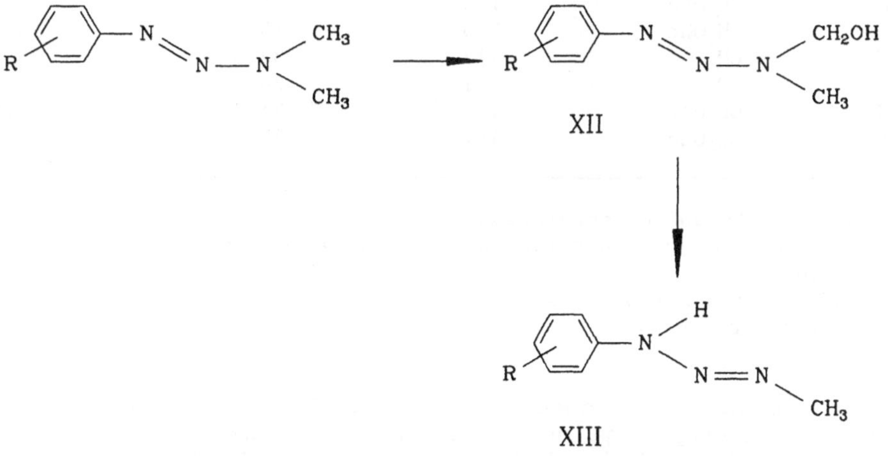

Figure 3.   Metabolism of 1-aryl-3,3-dimethyltriazene.

aryl-3-acetoxymethyl-3-methyltriazenes (XIV) and 1-aryl-3-methoxymethyl-3-methyltriazenes (XV) and 1-aryl-3-benzoyloxymethyl-3-methyltriazenes (XVI) (41). The antitumour activity of compounds of type XIV and XV against the PC6 plasmacytoma are shown in Table 8. The comparable activity of the acetoxymethyl and methoxymethyl derivatives of the 4-cyanophenyltriazenes and their dimethyl and monomethyl analogues is particularly interesting even though overall it is less than some of the others (41).

XIV	R = CH_2OOCCH_3		XVIII	R = CH_2SAryl
XV	R = CH_2OCH_3		XIX	R = $CH_2OAlkyl$
XVI	R = $CH_2OOCC_6H_5$		XX	R = CH_2SCH_3

Table 8. Activity of prodrugs of 1-aryl-3-hydroxymethyl-3-methyltriazenes against the PC6 tumour.

Substituents		$LD_{50}{}^A$ (mg/kg)	$ED_{90}{}^B$ (mg/kg)	TI^C
R	R^1			
$CONH_2$	CH_3	140	2.5	56
$CONH_2$	CH_2OAc	>200	17.5	>11.4
Ac	CH_2OAc	140	16.5	8.5
Ac	CH_2OMe	280	15	18.7
CN	CH_3	117	37.5	3.1
CN	H	71	20.5	3.5
CN	CH_2OAc	141	31	4.6
CN	CH_2OMe	141	33	4.3

A Lethal dose to 50% of population
B Dose to produce 90% reduction in volume of tumour compared with controls
C Therapeutic Index (LD_{50} ÷ ED_{90})
Data from reference 41

    This group also showed that the azido derivative (XVII) has similar activity against the PC6 tumour (42), and the possibility of preparing thio derivatives (XVIII) which may also be suitable prodrugs (43).

Iley and his colleagues have also reported similar alkyloxymethyl (XIX) and methylthiomethyl (XX) derivatives (44).

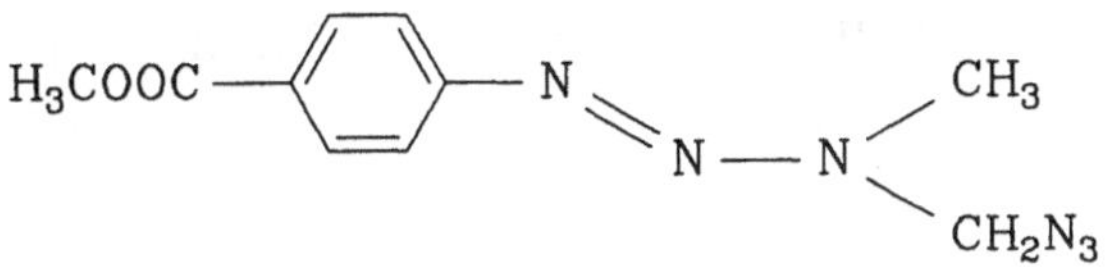

$$H_3COOC \longrightarrow \underset{N=N}{\overset{CH_3}{\underset{CH_2N_3}{\bigcirc}}} \qquad XVII$$

TRIAZENE N-OXIDES

An area of the triazenes which has been largely ignored is that of their N-oxides. This is surprising in view of the requirement of the 1-aryl-3,3-dimethyltriazenes for oxidative metabolism. Whilst various biological activities have been reported in the patent literature for the triazene N-oxides (45-47) it is only recently that this group of compounds have been investigated for potential antitumour activity. To date three types of triazene 1-oxides are the only triazene N-oxides to have been reported in the literature. These are the 1-aryl-3,3-dimethyltriazene 1-oxides (XXI), 3-aryl-1-methyltriazene 1-oxides (XXII) and the 3-aryl-1,3-dimethyltriazene 1-oxides (XXIII). Compounds of these three types have been prepared and their structure confirmed by crystallography, as detailed below.

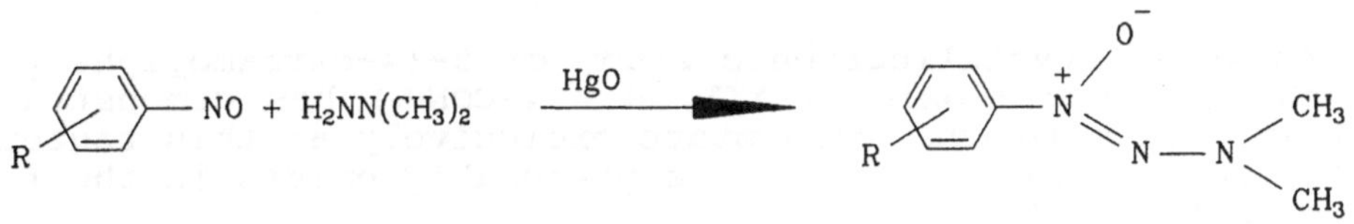

Figure 4.   Synthesis of 1-aryl-3,3-dimethyltriazene
1-oxides (XXI).

The 1-aryl-3,3-dimethyltriazene 1-oxides (XXI) are prepared, as shown in Figure 4, by the reaction of an arylnitroso compound (XXIV) with 1,1-dimethylhydrazine (XXV) in ethanol or ethyl acetate solution in the presence of mercuric oxide (46). The mercuric oxide inhibits the reduction of the nitroso compound. Confirmation of the presence of the oxygen at N^1 and of the structure as a whole has been obtained from an x-ray crystallographic study of 1-(4-carbamoylphenyl)-3,3-dimethyltriazene 1-oxide (XXI, R = 4-CONH$_2$) (53).
3-Aryl-1-methyltriazene 1-oxides (XXII) are prepared by the coupling reaction between the appropriate aryldiazonium salt (XXVI) and N-methylhydroxylamine (XXVII) under mildly basic aqueous conditions (18). Originally the location of the proton in compounds of this type, whether on oxygen or at N^3, was uncertain and they were usually described as 1-aryl-3-hydroxy-3-methyltriazenes (XXVII) (18, 49). However, ^{1}H and ^{13}C nuclear magnetic resonance spectroscopy has shown (50) that the isolable product from this reaction is of structure (XXII), irrespective of the electron-donating or electron-withdrawing properties of the aromatic ring. No evidence was seen for the existence of a 1-aryl-3-hydroxy-3-methyltriazene (XXVIII) in either deuterochloroform or hexadeuterodimethyl sulphoxide solution.

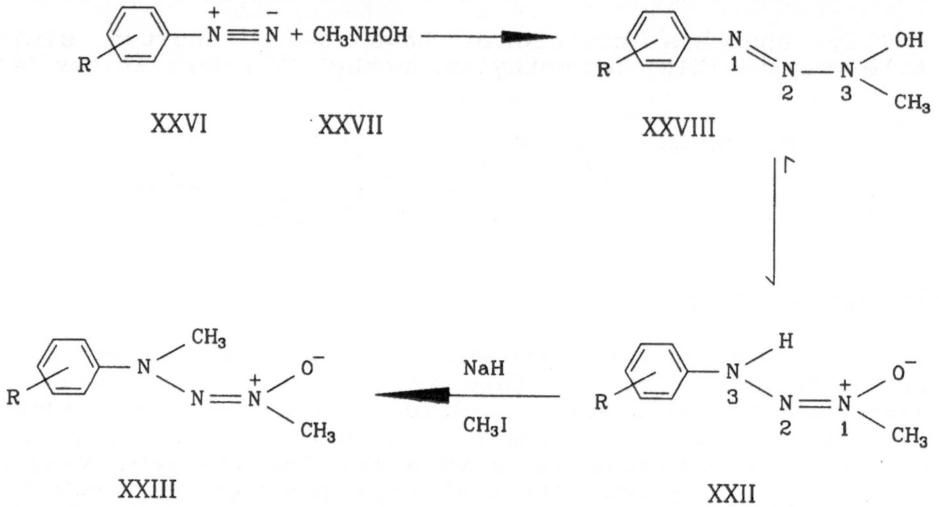

Figure 5. Synthesis of 3-aryl-1-methyltriazene
1-oxides (XXII) and 3-aryl-1,3-dimethyl-
triazene 1-oxides (XXIII).

An x-ray crystallographic study of 3-(4-carbamoylphenyl)-
1-methyltriazene 1-oxide (XXII, R = 4-$CONH_2$) has demonstrated
that it exists in the solid state exclusively as this tautomer
(51). The equilibrium therefore presumably exists in the form
shown in Figure 5.

Chemical evidence for the 3-aryl-1-methyltriazene 1-oxide
structure comes from its conversion to a 3-aryl-1,3-
dimethyltriazene 1-oxide (XXIII) (47) by sequential reaction
with sodium hydride and methyl iodide (Figure 5). Confirmation
of the location of the methyl group at N^3 and not on oxygen has
been obtained, in the case of 3-(4-carbamoylphenyl)-1,3-
dimethyltriazene 1-oxide (XXIII, R = 4-$CONH_2$) by x-ray
crystallography (52).

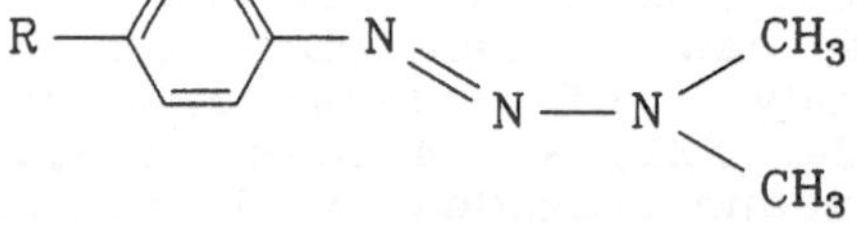

Table 9 shows the antitumour activity of a selection of the
three types of triazene 1-oxide against the ADJ/PC6A murine
plasmacytoma together with similar 1-aryl-3,3-dimethyltriazenes
(XXIX) and DTIC for comparison (53). The toxicity ranges (LD_{50})
of the various types of triazene 1-oxide are comparable, but
comparison of identically substituted compounds indicates
individual differences between the parent triazenes and their
N-oxides. The 3-aryl-1-methyltriazene 1-oxides and 3-aryl-1,3-

Table 9. Antitumour activity of some triazene 1-oxides against the Adj/PC6A mouse plasmacytoma.

Compound	Structure	R	Adj/PC6A mouse plasmacytoma		
			LD_{50}	ED_{90}	T.I.
1	XXI	H	35.5	21	1.69
2	XXI	$4\text{-}CONH_2$	141	18	7.83
3	XXI	$4\text{-}Cl$	71	-	
4	XXI	$4\text{-}N(CH_3)_2$	42	-	
5	XXII	$4\text{-}COOH$	>400	270	>1.48
6	XXII	$4\text{-}CONH_2$	330	15.5	21.3
7	XXII	$4\text{-}COOCH_3$	>100	-	
8	XXIII	$4\text{-}CONH_2$	285	195	1.46
9	XXIX	COOH	180	11	16.4
10	XXIX	$CONH_2$	140	2.5	56
DTIC			120	4	30

Adapted from reference 53.

dimethyltriazene 1-oxides are less toxic than the 1-aryl-3,3-dimethyltriazene 1-oxides which are of comparable toxicity with 1-aryl-3,3-dimethyltriazenes. Thus compounds which retain the aryl function at N^1 of the triazene chain have similar toxicity, whereas structural alterations resulting in it being at N^3 lead to reduced toxicity.

All of the N-oxides show reduced antitumour effectiveness (increased ED_{90}) as compared with the analogous 1-aryl-3,3-dimethyltriazenes.

As the antitumour effect of the 1-aryl-3,3-disubstituted triazenes is dependent on their ability to be converted chemically or metabolically to a 3-aryl-1-methyltriazene (18), the reduced antitumour effectiveness of the triazene 1-oxides may be a consequence of poor metabolic deoxygenation. No information on the metabolism of the triazene 1-oxides is available yet. The structural modifications made to the 1-aryl-3,3-methyltriazenes to produce their N-oxides will cause major changes in the partition and solubility properties of the compounds and hence their distribution, thus it is significant that activity is actually observed in this solid tumour system.

The activity of 3-aryl-1-methyltriazene 1-oxides is of particular interest. A knowledge of the requirement or otherwise for metabolic activation by loss of oxygen, of this class of compound, would indicate whether or not such compounds might be useful in the therapy of largely hypoxic tumours. In this situation, the anaerobic conditions might lead to ready reduction of the N-oxide to a cytotoxic monomethyltriazene metabolite.

Thus after almost 35 years of interest in the triazenes as antitumour agents, a new group of analogues, with the potential for activation by alternative routes, is available for detailed investigation.

REFERENCES

1. P. Griess, Uber eine neue Klasse organischer Verbindungen in denen Wasserstoff durch Stickstoff vertreten ist, *Ann. Chem.*, **121**: 257-280 (1862).
2. A. Baeyer and C.Jaeger, Ueber die Amide des Diazobenzols, *Ber.*, **8**: 148-151 (1875).
3. H. Druckrey, Specific carcinogenic and teratogenic effects of "indirect" alkylating methyl and ethyl compounds and their dependency on stages of oncogenic development, *Xenobiotica*, **3**: 217-303 (1973).
4. D.A. Clarke, R.K. Barclay, C.C. Stock and C.S. Rondestvedt, Triazenes as Inhibitors of Mouse Sarcoma 180, *Proc. Soc. Exptl. Biol. Med.*, **90**: 484-489 (1955).
5. J.H. Burchenal, M.K. Dagg, M. Beyer and C.C. Stock, Chemotherapy of Leukemia VII. Effect of substituted triazenes on transplanted mouse leukemia, *Proc. Soc. Expt. Biol. Med.*, **91**: 398-401 (1956).
6. H. Foerster and D. Steinhoff, Cytostatic *mono-* and *bis-*dialkyl triazenes, *South African Pat.* 69 04,895, *Chem. Abs.*, **73**, 55823 (1970).
7. Uniroyal Inc., Substituted triazenes as pre-emergent herbicides, *Brit. Pat.* 1,130,469, *Chem. Abs.*, **70**, 19788 (1969).
8. M. Mazza, G. Pagani, G. Calderara and L. Vicarini, Phytotoxic activity of triazene derivatives. I. 1-Phenyl-3,3-dialkyltriazenes phenyl-substituted with alkyl, halo or nitro groups, *Farmaco, Ed. Sci.*, **28**: 846-861 (1973).
9. Y.F. Shealy, R.F. Struck, L.B. Holum and J.A. Montgomery, Synthesis of potential anticancer agents. XXIX. 5-Diazoimidazole-4-carboxamide and 5-diazo-*v*-triazole-4-carboxamide, *J. Org. Chem.*, **26**: 2396-2401 (1961).
10. J.A. Montgomery, Experimental studies at Southern Research Institute with DTIC (NSC-45388), *Cancer Treat. Rept.*, **60**: 125-134 (1976).
11. Y.F. Shealy, C.A. Krauth and J.A. Montgomery, Imidazoles. I. Coupling reactions of 5-diazoimidazole-4-carboxamide (NSC-82196), *J. Org. Chem.*, **27**: 2150-2154 (1962).
12. Y.F. Shealy and C.A. Krauth, Complete inhibition of mouse leukaemia L1210 by 5(or 4)-[3,3-*bis*(2-chloroethyl)-1-triazeno]imidazole-4(or 5)-carboxamide, *Nature*, **210**: 208-209, (1966).
13. Y.F. Shealy, Triazenylimidazoles and related compounds, *Adv. in Med. Oncol. Res. Educ.*: Proc. Int. Cancer Congress 12th, Pergamon, Oxford, Vol 5: 49-58 (1978).
14. D.W. Miles and R.L. Souhami, Chemotherapy of metastatic disease, *Baillière's Clin. Oncol*, **1**: 551-573 (1987).
15. I. Kline, R.J. Woodman, M. Gang and J.M. Venditti, Effectiveness of antileukemia agents in mice inoculated with leukemia L1210 variants resistant to 5-(3,3-dimethyl-1-triazeno)imidazole-4-carboxamide (NSC-45388) or 5-[3,3-*bis*(2-chloroethyl)-1-triazeno]imidazole-4-carboxamide (NSC-82196), *Cancer Chemother. Repts*, **55**: 9-28 (1971).
16. Y.F. Shealy, C.A. Krauth, C.E. Opliger, H.W. Guin and W.R. Laster, Triazenes of phenylbutyric, hydrocinnamic, phenoxyacetic and benzoylglutamic acid derivatives, *J. Pharm. Sci.*, **60**: 1192-1198 (1971).

17. Y.F. Shealy, C.A. O'Dell, J.D. Clayton and C.A. Krauth, Benzene analogues of triazenoimidazoles, *J. Pharm. Sci.*, **60**: 1462-1468 (1971).

18. T.A. Connors, P.M. Goddard, K. Merai, W.C.J. Ross and D.E.V. Wilman, Tumour inhibitory triazenes: structural requirements for an active metabolite, *Biochem. Pharmacol.*, **25**: 241-246 (1976).

19. R.C.S. Audette, T.A. Connors, H.G. Mandel, K. Merai and W.C.J. Ross, Studies on the mechanism of action of the tumour inhibitory triazenes, *Biochem. Pharmacol.*, **22**: 1855-1864 (1973).

20. G.J. Hatheway, C. Hansch, K.H. Kim, S.R. Milstein, C.L. Schmidt and R.N. Smith, Antitumor 1-(X-aryl)-3,3-dialkyltriazenes. I. Quantitative structure-activity relationships vs. L1210 leukemia in mice, *J. Med. Chem.*, **21**: 563-574 (1978).

21. C. Hansch, G.J. Hatheway, F.R. Quinn and N. Greenberg, Antitumor 1-(X-aryl)-3,3-dialkyltriazenes. 2. On the role of correlation analysis in decision making in drug modification. Toxicity quantitative structure-activity relationships of 1-(X-phenyl)-3,3-dialkyltriazenes in mice, *J. Med. Chem.*, **21**: 574-577 (1978).

22. D. Farquhar and J. Benvenuto, 1-Aryl-3,3-dimethyltriazenes: potential central nervous system active analogues of 5-(3,3-dimethyl-1-triazeno)-imidazole-4-carboxamide (DTIC) *J. Med. Chem.*, **27**: 1723-1727 (1984).

23. V.A. Levin, Relationship of octanol/water partition coefficient and molecular weight to rat brain capillary permeability, *J. Med. Chem.*, **23**: 682-684 (1980).

24. D.E.V. Wilman, N.J. Bradley and S.G. Richardson, Astrocytoma xenografts in the choice of a second-generation triazene, *Brit. J. Cancer*, **50**: 277 (1984).

25. G.J. Atwell, B.F. Cain and W.A. Denny, Potential antitumor agents. 22. Latentiated congeners of the 4'-(9-acridinyl-amino)methanesulfonanilides, *J. Med. Chem.*, **20**: 520-526 (1977).

26. G. Abel, T.A. Connors, P. Goddard, H. Hoellinger, N.-H. Nam, L. Pichat, W.C.J. Ross and D.E.V. Wilman, Cytotoxic sulphonamides designed for selective deposition in malignant tissues, *Europ. J. Cancer*, **11**: 787-793 (1975).

27. A.J. Lin and T.L. Loo, Synthesis and antitumor activity of halogen-substituted 4-(3,3-dimethyl-1-triazeno)quinolines, *J. Med. Chem.*, **21**: 268-272 (1978).

28. B.S. Ashby, pH Studies in human malignant tumours, *Lancet*, **ii**: 312-315 (1966).

29. C.D. Stevens, M.A. Wagner, P.M. Quinlin and A.M. Kock, Localization of sulfapyrazine in cancer tissue on glucose injection, *Cancer Res.*, **12**: 634-639 (1952).

30. G. Abel, T.A. Connors, W.C.J. Ross, N.-H. Nam, H. Hoellinger and L. Pichat, The selective concentration of sulphadiazine and related compounds in malignant tissue, *Eur. J. Cancer*, **9**: 49-54 (1973).

31. N. Calvert, T.A. Connors and W.C.J. Ross, Aryl-2-halogenoalkylamines XXV. Derivatives of sulphanilamide designed for selective deposition in neoplastic tissue. *Europ. J. Cancer*, **4**: 627-636 (1968).

32. S.M. Sieber and R.A. Adamson, Selection of carriers for alkylating moieties to increase their antitumor specificity, *Cancer Treatment Rep.*, **60**: 217-219 (1976).

33. R.E. Counsell, P. Pocha, J.O. Morales and W.H. Beierwaltes, Tumor localizing agents III. Radioiodinated quinoline derivatives, *J. Pharm. Sci.*, **56**: 1042-1044 (1967).

34. E. Bonmassar, A. Bonmassar, S. Vadlamudi and A. Goldin, Immunological alteration of leukemic cells *in vivo* after treatment with an antitumor drug, *Proc. Natl. Acad. Sci. U.S.*, **66**: 1089-1095 (1970).

35. F.A. Schmid and D.J. Hutchison, Decrease in oncogenic potential of L1210 leukemia by triazenes, *Cancer Res.*, **33**: 2161-2165 (1973).

36. A. Nicolin, F. Spreafico, E. Bonmassar and A. Goldin, Antigenic changes of L5178Y lymphoma after treatment with 5-(3,3-dimethyl-1-triazeno)imidazole-4-carboxamide *in vivo*, *J. Nat. Cancer Inst.*, **56**: 89-93 (1976).

37. A. Nicolin, M. Cavalli, A. Missiroli and A. Goldin, Immunogenicity induced *in vivo* by DIC in relatively non-immunogenic leukemias, *Europ. J. Cancer*, **13**: 235-239 (1977).

38. T. Giraldi, A.M. Guarino, C. Nisi and G. Sava, Antitumor and antimetastatic effects of benzenoid triazenes in mice bearing Lewis lung carcinoma, *Pharmacol. Res. Commun.*, **12**: 1-11 (1980).

39. D.J. Kohlsmith, K. Vaughan and S.J. Luner, Triazene metabolism. III. *In vitro* cytotoxicity towards M21 cells and *in vivo* antitumor activity of the proposed metabolites of the antitumor 1-aryl-3,3-dimethyltriazenes, *Can. J. Physiol. Pharmacol.*, **62**: 396-402 (1984).

40. K. Vaughan, Y. Tang, G. Llanos, J.K. Horton, R.J. Simmonds, J.A. Hickman and M.F.G. Stevens, Studies of the mode of action of anti-tumor triazenes and triazines. 6. 1-Aryl-3-hydroxymethyl-1-methyltriazenes: synthesis, chemistry and anti-tumor properties, *J. Med. Chem.*, **27**: 357-363 (1984).

41. L.M. Cameron, R.J. LaFrance, C.M. Hemes, K. Vaughan, R. Rajaraman, D.C. Chubb and P.M. Goddard, Triazene metabolism IV. Derivatives of hydroxymethyltriazenes: potential prodrugs for the active metabolites of the antitumour triazene, DTIC, *Anti-Cancer Drug Des.*, **1**: 27-36, (1985).

42. K. Vaughan, K.U.K. Gamage Nicholas, R.D. Singer, M. Roy and N.W. Gibson, Triazene metabolism. VI. 3-Azidomethyl-3-alkyl-1-aryltriazenes, a new class of antitumour triazene with potential pro-drug applications, *Anti-Cancer Drug Des.*, **2**: 279-287 (1987).

43. K. Vaughan, H.W. Manning, M.P. Merrin and D.L. Hooper, Open chain nitrogen compounds. Part XIII. 1-Aryl-3-arylthiomethyl-3-methyltriazenes and 3-(arylazo)-1,3-thiazolidines, *Can. J. Chem.*, **66**: 2487-2491 (1988).

44. J. Iley, E. Rosa and L. Fernandes, Triazene drug metabolites. Part 5. A simple direct synthesis of 3-alkoxymethyl- and 3-alkylthiomethyl-1-aryl-3-alkyltriazenes from 1-aryl-3-hydroxymethyl-3-alkyltriazenes, *J. Chem. Research (S)*, 264-265 (1987).

45. F.K. Hess, P.A. Stewart, G. Possanza and K. Freter, Immunosupressive 1-phenyl-3-hydroxy-3-methyltriazenes, *Ger. Patent*, 2,208,368, *Chem. Abs.*, **79**: 146271 (1973).

46. J.L. Miesel, 3,3-Dialkyl-1-(substitutedphenyl)triazene-1-oxides, *U.S. Patent*, 3,989,680 (1976).

47. J.L. Miesel, 3-Aryltriazene 1-oxides for treating inflammatory diseases, *U.S. Patent*, 3,962,434 (1976).

48. S. Neidle and D.E.V. Wilman, Unpublished data.

- 49. K. Freter, F. Hess and K. Grozinger, Acylierung von Hydroxy-phenyl-triazenes, eine neuartige Umlagerung, *Ann. Chem.*, 811–820 (1973).
- 50. A.G. Giumanini, L. Lassiani, C. Nisi, A. Petric and B. Starovnik, The structure of aryldialkyltriazene *N*-oxides, *Bull. Chem. Soc. Jpn.*, **56**: 1887–1888 (1983).
- 51. R. Kuroda and D.E.V. Wilman, 3-(4-Carbamoylphenyl)-1-methyltriazene 1-oxide, *Acta Cryst.*, **C41**: 1543–1545 (1986).
- 52. S. Neidle, G.D. Webster, R. Kuroda and D.E.V. Wilman, Synthesis and structure of 3-(4-carbamoylphenyl)-1,3-dimethyltriazene 1-oxide, *Acta Cryst.*, **C43**: 674–676 (1987).
- 53. D.E.V. Wilman and P.M. Goddard, Triazene *N*-oxides. 2. Synthesis and antitumour activity of some triazene 1-oxides, *Prog. in Pharmacol. Clin. Pharmacol.*, in press.

ANTIMESTASTATIC ACTION OF TRIAZENE DERIVATIVES

Tullio Giraldi #, Laura Perissin, Sonia Zorzet
and Valentina Rapozzi

Istituto di Farmacologia Università di Trieste
I-34100 Trieste, Italy

INTRODUCTION

Dimethyltriazenes have gained, since the original observation in 1962 by Shealy et al. on the antitumor action of DTIC in mice (1), a definite position among antineoplastic drugs because of the established activity of DTIC against human malignant melanoma, as well as against other human malignancies when used in combination with other antitumor drugs. Substituted benzenoid dimethyltriazenes have also been thoroughly examined in experimental animal tumor systems, and selected compound(s) are currently under clinical trials as second generation substitutes for DTIC (see other chapters in this volume).

The major problem encountered in the clinical treatment of solid malignant tumors is constituted by the (late) appearance of systemic metastasis. In this connection, the observation by Heyes et al. in 1974 , reporting that the imidazole triazene derivative BRL 51308 could reduce in mice bearing Lewis lung carcinoma the formation of lung metastasis independently from primary tumor (2), appears of particular interest. Further investigations have been subsequently performed on the specific action of triazene derivatives on metastasis, showing that dimethyltriazenes possess, in addition to several biological and pharmacological properties, a rather selective antimetastatic action, dependent on their chemical structure.

The aim of this paper is thus that to illustrate the antimetastatic action of triazene derivatives; the literature available on the effects of triazenes on metastasis of solid malignant tumors in mice will be reviewed. In order to avoid repetitions, reference will be made only to work which is not dealt with in detail in other chapters of this volume, which should be consulted for specific detailed infor-

---

Present address: Cattedra di Farmacologia, Istituto di Biologia, Facoltà di Medicina, Università di Udine, I-33100 Udine Italy.

EFFECTS OF IMIDAZOLE AND BENZENOID DIMETHYLTRIAZENES ON SPONTANEOUS METASTASIS OF SOLID MALIGNANT TUMORS

The effects of dimethyltriazenes on spontaneous hematogenous metastasis of solid tumors are typically observed when mice implanted s.c. with Lewis lung carcinoma are treated i.p. daily during the two weeks following tumor implantation. Using these conditions and equitoxic drug dosages, including maximum tolerated doses for the treatment schedule employed, DTIC and BRL 51308 cause only a slight decrease in the growth of s.c. primary tumor, whereas $DM-CONH_2$ is devoid of significant effects. On the contrary, all of the drugs markedly inhibit in a dose-dependent way the development of spontaneous lung metastasis, causing at the highest dose level the absence of macroscopically detectable tumor lung nodules in a large proportion of the treated animals (3). Similar effects, but remarkably more pronounced on metastasis, are obtained with DM-COOK (4). The formation of spontaneous lung metastasis appears to be inhibited by a specific mechanism different from cytotoxicity of the tested drugs for tumor cells. This is indicated by insignificant effects of the treatment with the tested triazenes, unlike cytotoxic drugs such as cyclophosphamide, on s.c. tumors (3,4). Further evidence in this connection is provided by the inactivity of DM-COOK, $DM-NO_2$ and $DM-CH_3$ on the development of pulmonary tumor colonies obtained by i.v. injection of tumor cells (4). These findings are also supported by the examination of the *fractional incorporation* of ^3H-labeled thymidine in tumor cells, which indicates the absence of cytotoxicity of the tested triazenes for tumor cells localized in the lungs (3,4). The antimetastatic action of DM-COOK is exerted, independently from its inhibitory effects on primary tumor, also in mice bearing B16 melanoma (5), M5076/73A ovarian reticular cell sarcoma (6) and MCa mammary carcinoma (7).

STRUCTURE ACTIVITY RELATIONSHIPS FOR THE ANTIMETASTATIC EFFECTS OF ARYL-DIMETHYLTRIAZENES

Quantitative structure-activity relationships for p-substitute aryldimethyltriazenes

Quantitative structure-activity relationships (*QSAR*) for the differential effects on spontaneous metastasis in comparison with s.c. primary tumor have been examined for a series of *p*-substituted phenyldimethyltriazenes in mice bearing Lewis lung carcinoma. Compounds with substituents conferring to the molecule a wide range of electronic and/or hydrophobic constants, have been examined; the expression of drug activity as *log T/C*, rather than *log 1/C*, has been found to produce much more meaningful results. The *QSAR* analysis of the data obtained indicate a limited dependency of the effects of the tested drugs on the electronic properties of the substituents, with increased activity for electron releasing substituents; hydrophilicity (*log p*) appears to outweigh the relevance of electronic effects. Moreover, the activity on primary tumor reveals a maximum for a *log p* value of about 2, which corresponds to a minimum of activity on metastasis. The effects on metastasis strongly depend on *log p*, and are more pronounced for highly lipophilic or hydrophilic substituents. These findings quantitatively indicate the dissociation of the effects of phenyldimethyltriazenes on primary tumor from those on metastasis, and confirm the optimal properties of (hydrosoluble) DM-COOK (8).

Isomerism of aryldimethyltriazenes

The effects of the *ortho* and *para* isomers of aryldimethyltriazeno carboxylic and benzensulfonic acids and amides have also been examined in mice bearing Lewis lung carcinoma. At equitoxic dosages, and even at maximum tolerated levels, the tested compounds reduce only marginally the growth of s.c. primary tumors; on the contrary, the formation of spontaneous lung metastasis is strongly inhibited. The compounds showing the greatest activity are the hydrosoluble salts of the *ortho* and *para* carboxylic acid and the *para* sulfonamido derivative; the activity of the hydrosoluble sodium salt of the *ortho* benzensulfonic acid is significantly less pronounced than that of the other compounds (9). These results further confirm for substituted aryldimethyltriazenes the dissociation of the effects on metastasis from those on primary tumor, as well as the optimal activity of DM-COOK.

BIOTRANSFORMATION AND ANTIMETASTATIC ACTION OF DIMETHYLTRIAZENES

A considerable amount of evidence has accumulated, showing that dimethyltriazenes undergo significant biotransformation either by proton catalyzed chemical hydrolysis or by enzyme catalyzed hepatic metabolism (10). The mechanism(s) of these transformation(s), and their role in the cytotoxic action of dimethyltriazenes is dealt with in detail in other parts of this volume; this section will illustrate only the data available concerning the role of biotransformation for the antimetastatic action of aryldimethyltriazenes.

Hydrolysis of aryldimethyltriazenes to aryldiazonium cations

The role of aryldiazonium cations for determining the effects of administered aryldimethyltriazenes can be investigated using derivatives with electron withdrawing or releasing substituents. These compounds have different half-life time ($T_{1/2}$) of hydrolysis, and can generate during the time extension of *in vivo* experiments variable amounts of diazonium ions, depending on the kinetic parameters of the hydrolysis reaction (11). In general terms, the electronic effects of the substituents are largely outweighed by their contribution to lipophilicity (8). However, significant differences in biological activity are displayed by compounds with substituents causing different electronic effects. DM-NO$_2$, DM-CONH$_2$ and DM-CH$_3$ have a $T_{1/2}$ for hydrolysis to aryldiazonium cations widely ranging from 160 days for DM-NO$_2$, to 32 min for DM-CH$_3$, that for DM-CONH$_2$ being of 7.15 days (12). When these compounds are tested at equitoxic dosages, including maximum tolerated ones, in mice bearing

Lewis lung carcinoma, the effects on s.c. primary tumor range from marked inhibition (DM-NO$_2$) to inactivity (DM-CH$_3$), those of DM-CONH$_2$ falling in between. At the same time, the formation of lung metastasis is strongly inhibited, substantially to the same degree, by all of the three compounds (12). Incidentally, $T_{1/2}$ for hydrolysis to diazonium of DM-COOK, which has the most pronounced antimetastatic activity among dimethyltriazenes, is 2.0 days (13). On the other hand, when the corresponding p-substituted phenyldiazonium tetrafluoborate salts are administered in vivo to mice implanted with Lewis lung carcinoma, no effect on metastasis is observed even at maximum tolerated doses; at the same time, the growth of primary s.c. tumor is only marginally inhibited (14).

These findings altogether indicate that the systemic extracellular presence of aryldiazonium cations does not result in antimetastatic effects, and indicate that the antimetastatic action of dimethyltriazenes is caused also by compounds, such as DM-NO$_2$, which can not generate appreciable amounts of diazonium cations throughout the duration of in vivo experiments.

Hepatic metabolism of aryldimethyltriazenes

The microsomal oxidative N-demethylation of dimethyltriazenes to the corresponding monomethyltriazenes is a further metabolic pathway which has been investigated, and suggested to be involved for the cytotoxic action of dimethyltriazenes (15,16). The direct measurement of this biotransformation in vivo is difficult, since arylmonomethyltriazenes have an extremely short $T_{1/2}$ for decomposition in aqueous solution at physiological pH (see the chapter by K. Vaughan in this volume). Data obtained in vitro are available, showing that substituted aryldimethyltriazenes undergo N-demethylation to different degrees, depending on the nature of the substituents (17,18). Additional evidence indicating that other metabolic biotransformations may also occur on dimethyltriazenes, and may be of relevance for their antitumor activity, is available but still require definitive clarification (19,20).

In order to investigate the relevance of demethylation for the antimetastatic action of dimethyltriazenes, the effects obtained in vivo with DM-COOK, DM-NO$_2$ and DM-CH$_3$ against Lewis lung carcinoma have been related with in vitro kinetic parameters measured for N-demethylation. All of the three drugs cause effects of similar magnitude, consisting of a marginal inhibition of primary s.c. tumor accompanied by pronounced antimetastatic effects. At the same time in vitro N-demethylation of DM-COOK is very limited and not measurable; that of DM-CH$_3$ has $V_{max}=2.7$ and $K_m=0.06$, whereas for DM-NO$_2$ V_{max} and K_m are equal to 10.6 and 0.11 respectively (18). These results indicate, at least for the three dimethyltriazenes examined, a lack of correlation between the K_m and V_{max} for N-demethylation and antimetastatic activity. In a further series of experiments, the animals have been treated with an inducer (phenobarbital, PB) or an inhibitor (carbon tetrachloride, CCl$_4$) of hepatic drug metabolism. The treatment with PB or CCl$_4$ at the dosages and schedules employed proved to be effective in markedly modifying the N-demethylation of the three dimethyltriazenes examined, as had been determined in vitro. Yet, no unambiguous increase by PB, or decrease by CCl₄, which should be theoretically expected if metabolic transformation of dimethyltriazenes through demethylation (and/or other pathways susceptible to be influenced by PB or CCl₄) were required for their action, was observed either on primary tumor or metastasis (18).

The importance of N-demethylation has also been determined by direct administration of 1-*p*-nitro, 1-*p*-carboxamido and 1-*p*-tolyl-3-methyltriazene (MM-NO₂, MM-CONH₂ and MM-CH₃) to mice bearing Lewis lung carcinoma, in comparison with the parent dimethyl-derivative DM-NO₂, DM-CONH₂ and DM-CH₃. Maximum tolerated doses of monomethyltriazenes are smaller than those of dimethyltriazenes, and the effects on metastasis are less pronounced than those of dimethyltriazenes. The monomethyltriazenes MM-CONH₂ and MM-CH₃ retain the marginal activity on primary s.c. tumor of their parent dimethyl derivatives, and MM-NO₂ is devoid of the significant activity on primary tumor displayed by DM-NO₂ (12).

Conclusions

The data presented indicate a lack of correlation between biological effects (antitumor and antimetastatic activity *in vivo* against Lewis lung carcinoma) and the capacity of dimethyltriazenes to generate aryldiazonium cations. A similar lack of correlation is found comparing kinetic parameters for (*in vitro*) demethylation with (*in vivo*) biological activity. This finding is in agreement with the results obtained with the administration of an inducer or an inhibitor of microsomal drug metabolism.

These results are also supported by the inactivity observed after *in vivo* administration of aryldiazonium cations, and by the limited activity of *in vivo* administered monomethyltriazenes. These findings appear to indicate that dimethyltriazenes need a direct interaction with tumor cells without undergoing biotransformation in order to exert their antimetastatic action. No information is available on their possible biotransformation within tumor cells, and the relevance of this event has not been ascertained so far. At the same time no clear indication appears for the relevance of biotransformation for the cytotoxic action of dimethyltriazenes on primary s.c. Lewis lung carcinoma, which cannot be further analyzed because of the natural unresponsiveness to triazenes of primary implants of this tumor.

MECHANISM OF THE ANTIMETASTATIC ACTION OF DIMETHYLTRIAZENES

In order to exert their antimetastatic action, dimethyltriazenes require the direct interaction of the drug with the primary tumor. Indeed, DM-COOK, DM-NO₂ and DM-CH₃ are devoid of effects on artificial metastasis obtained by i.v. inoculation of Lewis lung carcinoma tumor cells (4,14). Moreover, the antimetastatic action of DM-COOK is not observed when normal mice are pretreated with the drug, and the tumor is implanted s.c. at the end of drug treatment (4).

Effects of DTIC and DM-COOK on hemostasis

A sufficient number of thrombocytes and an active blood clotting system have been shown to be required for tumor cell dissemination (21,22); drugs acting on platelets and blood coagulation correspondingly inhibit tumor metastasis (23,24). The effects of DTIC and DM-COOK on hemostasis have thus been investigated; the parameters considered are the number of platelets and their aggregability, prothrombin and partial thromboplastin times, plasma fibrinogen concentration and tumor cell procoagulant activity. Slight variations in the parameters considered are caused by DTIC and DM-COOK, when administered at dosages and treatment schedules effective as antimetastatic, in tumor bearing mice as compared with drug untreated tumor bearing controls (25). The magnitude of the effects on hemostasis of DTIC and DM-COOK is limited, and insufficient to significantly inhibit metastasis, as found in other investigations on metastasis in relation to blood coagulation (26).

Effects of DTIC and DM-COOK on tumor proteinases

The capacity of solid malignant tumors to produce hematogenous systemic metastasis has been shown to be associated with the secretion by tumor cells of neutral proteinases, such as plasminogen activator (27), cathepsin B (28) and collagenases (29). These enzymes may be held responsible for detachment of disseminating cells from tumor parenchyma, for lysis of vascular basement membrane components during the phases of tumor cell intravasation and extravasation, and for degradation of extracellular matrix components before intravasation and during metastatic tumor cell lodgement in the target organ (30). These findings are supported by the antimetastatic action observed after *in vivo* administration of neutral proteinase inhibitors of natural and synthetic origin (31,32), including diazoderivatives of aminoacids (33-35). The possibility that the antimetastatic action of dimethyltriazenes might be exerted by means of the inhibition of proteinases associated with tumor cells has been consequently investigated.

The activity of cathepsin B and plasminogen activator has been determined in homogenates of *ex vivo* tissue samples of two lines of Lewis lung carcinoma selected for high (line M1087) or low (line BM21548) potential to spontaneously metastasize, and of B16 melanoma; cathepsin B content has been determined also in cell cultures *in vitro* of Lewis lung carcinoma line C108. The effects of DM-COOK have been compared with those caused by the reference antimetastatic drug ICRF 159 (36), and by the cytotoxic antitumor drug CCNU. All these drugs strongly inhibit metastasis in mice bearing Lewis lung carcinoma. The action of DM-COOK occurs only on spontaneous metastasis (selective antimetastatic) (4), and that of CCNU is also markedly pronounced on primary tumor (cytotoxic) (37); ICRF 159 has a mixed antimetastatic and cytotoxic activity (38). No correlation appeares between metastatic potential and the measured levels of cathepsin B and plasminogen activator. Similarly, although a certain degree of inhibition, more evident on cathepsin B, is observed in tumor samples obtained from animals treated *in vivo*, no correlation appeares between enzyme inhibition and antimetastatic action of the tested drugs (39). This investigation was subsequently extended to a broader panel of tumors, including B16 melanoma and MCa mammary carcinoma, and to include the cytotoxic antitumor drugs GANU, cisplatin and cyclophosphamide. Again, a marginal inhibition was observed for some drug-tumor combinations, but no meaningful pattern of inhibition either based on tumor metastatic potential or on drug's mechanism of action could be recognized (40).

An enzymatic activity which has been also related with metastatic behavior of tumor cells is a trypsin-like neutral proteinase associated with tumor cell surface (41). For Ehrlich ascites tumor cells, the *in vitro* incubation with millimolar concentrations of DM-COOK results in the total inhibition of this enzymatic activity, as well as of tumor cell induced lysis of erythrocytes (42). However, when the levels of trypsin-like neutral proteinase are assayed in two Lewis lung carcinoma lines with different metastatic potential, no correlation appeares between enzyme levels and metastatic capacity. Moreover, the antimetastatic action of DM-COOK is not accompanied by enzyme inhibition, and among the other drug tested (ICRF 159, CCNU, GANU and cisplatin) only CCNU causes significant inhibition (43).

These findings seem to rule out the inhibition of plasminogen activator, cathepsin B, or trypsin-like neutral proteinase associated with tumor cell membrane as the main target of DM-COOK antimetastatic activity. In analyzing these data, it has however to be noted that the assay of tumor proteinases, particularly in *ex vivo* drug treated tumor specimens, offers considerable inherent methodological problems (44), and that the possible inhibition of other proteinases involved in tumor metastasis, such as collagenases, does not appear to have been investigated so far.

Morphological and cytofluorimetric analysis of the effects of DTIC and DM-COOK

The cytological and histological characteristics of the Lewis lung carcinoma lines with high (M1087) and low (BM21548) potential to spontaneously metastasize have been analyzed. Metastatic potential appears to be related with parenchymal organization of the primary tumors, since large haemorrhagic areas containing detached tumor cells and the absence of endothelial capillaries are observed only in the line with high metastatic potential; the cytological characteristics of the cells of both lines are similar. After *in vivo* treatment with DTIC, DM-COOK or ICRF 159, the histological appearance of the line with high metastatic potential becomes similar to that of the line with low metastatic potential (45). These data indicate that the dimethyltriazenes DTIC and DM-COOK can act on primary tumor, causing specific morphological alterations which have been associated with a decrease in tumor cell intravasation during studies performed with ICRF 159 (46). They also suggest that this action might be of relevance for the clinical effects of DTIC.

This finding is supported by the measurement of the number of circulating tumor cells, performed by means of flow cytometry on the blood of mice bearing Lewis lung carcinoma. This technique was developed for overcoming the difficulties encountered in similar approaches, does not require either concentration of nucleated cells or other processing of blood samples, and is applicable to aneuploid tumors such as Lewis lung carcinoma. After treatment with DTIC, DM-COOK or ICRF 159, the number of circulating tumor cells is markedly reduced. This reduction is particularly

evident for DM-COOK, which abolishes the presence of circulating tumor cells at 8 and 12 days after tumor implantation; the reduction at 15 days, and that caused by the DTIC
and ICRF 159 at the three time periods examined falls in the
range 72-83 % (26).

These results are in agreement with those obtained by
the morphological analysis of the effects of drug treatment
on primary tumors in showing that the treatment with dimethyltriazenes prevents the entry of metastasizing tumor cells
into the blood stream. They also further support the view
that this action of DTIC may participate into its clinical
action.

Responsiveness of tumor lines with different metastatic potential to DTIC and DM-COOK

Solid malignant tumors have been shown to be constituted by heterogeneous populations of tumor cells endowed with
various biological properties (47,48), including different
metastatic potential (49). It is hence theoretically possible that dimethyltriazenes may cause their antimetastatic
action, which is not accompanied by inhibition of primary
tumor growth, by means of the selection of pre-existing
tumor cells clones with reduced metastatic potential among
those originally present before treatment in the heterogeneous parental tumor.

This possibility has been explored by the examination
of the effects of dimethyltriazenes on cell lines of Lewis
lung carcinoma, clonally selected for different potential to
spontaneously metastasize to the lungs (50). Tumor lines
BC215 and C108, which have a low and high metastatic potential *in vivo* respectively, have been treated *in vitro* with
DTIC and DM-COOK in the presence of a drug metabolizing
system. The obtained survival curves show a greater chemosensitivity of the C108 line to DM-COOK, whereas both lines
respond equally to DTIC. The effects of DM-COOK are observed
only in the presence of the metabolic activation system,
whereas DTIC causes pronounced effects also without metabolic biotransformation. Moreover, the mechanism of the cytotoxic action of the two triazenes appears to be exerted with
a different mechanism, since survival curves of tumor cells
show an exponential trend for DM-COOK and a threshold exponential trend for DTIC (51). The effects of DTIC and DM-COOK
have been also examined in mice bearing *in vivo* stabilized
Lewis lung carcinoma lines with low (BM21548) and high
(M1087) metastatic potential. DTIC causes a similar depression of primary tumor growth and artificial metastasis for
both tumor cell lines. On line BM21548 (low metastatic), the
inhibitory effects of DM-COOK are remarkably pronounced on
primary tumor and artificial metastasis, whereas they are of
marginal magnitude on line M1087. On the other hand, the
effects of both drugs on spontaneous metastasis are more
pronounced on line M1087 (high metastatic) (52).

The data presented indicate that the tumor line with
lower metastatic potential is more sensitive to the cytotoxic action of the tested dimethyltriazenes, whereas the tumor
line with higher metastatic potential is more responsive to
the antimetastatic action of the same drugs. These findings
therefore lead to exclude the possibility of a clonal selection of tumor cell populations with reduced metastatic
potential as the mechanism of the antimetastatic action of

dimethyltriazenes, and further confirm that the antimetastatic properties of these compounds do not depend on their cytotoxic effects.

Effects of long term treatment with DTIC and DM-COOK on tumor metastatic potential

The treatment of mice bearing transplantable leukemias for several transplant generations results in the induction of the appearance of strong immunogenicity, with loss of transplantability in syngeneic immunocompetent animals (chemical xenogenization) (53). The treatment of mice bearing solid malignant tumors with dimethyltriazenes for several transplant generations has been consequently examined, in terms of metastatic potential, local growth and for the possible occurrence of chemical xenogenization.

Mice bearing Lewis lung carcinoma have been treated for a variable number of tumor transplant generations with DTIC or DM-COOK. The metastatic potential, and the local i.m. growth of primary tumor, have been subsequently determined for the treated tumor cells by further i.m. transplantation in syngeneic normal recipient mice. A significant reduction in metastatic potential is observed after one transplant generation, with a magnitude slightly increasing when drug treatment is extended to three transplant generations; the reduction in metastatic potential is retained for three transplant generations of observation without treatment. In these conditions, no significant alteration in the growth of the primary i.m. tumor accompanies the reduction in metastatic potential. When the normal recipient mice are immunosuppressed by cyclophosphamide before tumor implantation, the reduction in the metastatic potential of tumors previously treated with dimethyltriazenes is less pronounced. The difference of metastatic potential in immunocompetent vs. immunodepressed hosts tends to level off with an increasing number of transplant generations of observation (54). Results qualitatively similar to those obtained with Lewis lung carcinoma, but slightly less pronounced quantitatively, are obtained with MCa mouse mammary carcinoma (55). When the number of transplant generations of treatment with DTIC is increased to ten, and the treated tumor is transplanted into normal hosts, i.m. growth is retained but metastatic potential is completely suppressed; immunosuppression of recipient hosts results in a limited appearance of lung metastasis (56).

These data indicate that the dimethyltriazenes DTIC and DM-COOK cause in tumor cells a substantially stable reduction of metastatic potential. The phenomenon of chemical xenogenization does not appear to be involved in the same clear way as observed with mouse transplantable leukemias, although a limited participation may be operative. The *in vitro* treatment of B16 melanoma cells with MM-COOK has been found to reduce the tumorigenic and metastatic properties, as determined by *in vitro* and *in vivo* analysis of the treated tumor cells. These effects are associated with an increase in the T cell dependent immune responses of the host against the treated tumor cells (57). The relevance of this result for the *in vivo* antitumor and antimetastatic action of DM-COOK appears questionable, considering that hepatic demethylation determined for this dimethyltriazene is negligible (18,58). The mechanism by which dimethyltriazenes cause the persistent reduction of metastatic potential does not seem clarified at present, and might involve somatic mutations induced by the drugs as well as other epigenetic actions.

THERAPEUTIC POTENTIAL AND TOXICITY OF THE ADJUVANT ANTIMETA-
STATIC TREATMENT WITH DIMETHYLTRIAZENES COMBINED WITH SURGI-
CAL REMOVAL OF PRIMARY TUMOR

The actual protocols for the clinical treatment of solid malignant tumors are mainly based upon the surgical removal of primary tumor combined with the administration of cytotoxic antitumor drugs. On this basis, and also consider-ing that dimethyltriazenes have a marginal cytotoxicity against Lewis lung carcinoma and B16 melanoma, the effects of DTIC and aryldimethyltriazenes have been examined in mice bearing solid metastasizing tumors when used as adjuvants to surgical treatment.

Increase in survival time

When B16 melanoma is implanted i.m. into the calf of the hind leg and the tumor bearing leg is amputated on day 9 from tumor implantation, all of the mice die within 4 months because of systemic metastasis. The daily treatment with DM-COOK or DM-CH$_3$ on days 1-8 combined with surgery causes a similar proportion (36-43%) of long term survivors (cured mice, survival time greater than 4 months). The proportion of cures is slightly reduced when treatment is performed on days 7-9 (23-35%), and is limited to 10-11% for a single pre-operative administration (59). The effects of DM-COOK have been compared in analogous experiments, where the leg implanted with B16 melanoma was amputated on day 11. Both drugs were similarly effective, causing about 60% cures either when administered on days 1-11 or 5-11. The treatment on days 7-14 of more advanced tumors removed on day 14 is less effective, causing 25-40% cures (5). The survival time of uncured mice is significantly increased in both investi-gations (5,59), and the post-operative treatment with DM-COOK is ineffective (5). The effects of the treatment with DM-COOK combined with surgery are not limited to B16 melano-ma. On Lewis lung carcinoma, the amputation of the tumor bearing leg on day 9 is not curative; the daily treatment on days 1-8 or 7-9 combined with surgery causes cures in about 40% of the animals (60,61). In mice amputated on day 12, the effects of DM-COOK are markedly more pronounced than those of DTIC (7). Similar results have been obtained with M5 ovarian reticular cell sarcoma and MCa mammary carcinoma, with percentages of cures of 14-40 or 25 respectively, and significant increases in the life-span of uncured animals (6,62).

Mechanism of the increase in survival time

The mechanism of the antimetastatic adjuvant action of the pre- and intra-operative treatment with dimethyltria-zenes combined with surgery, appears to be complex. Indeed, the reduction in metastasis determined either macroscopical-ly or histologically in mice bearing Lewis lung carcinoma can not wholly account for the increase in life-span ob-

served for the combination with surgery (45,61). The available experimental evidence indicates that natural resistance
factors of the host are crucial for the total eradication of
(micro)metastasis after treatment with triazenes and surgery
(61,63). The re-implantation of Lewis lung carcinoma in mice
cured by DM-COOK plus surgery results in significantly
smaller numbers of spontaneous and artificial metastasis as
compared with non-tumored mice treated with DM-COOK and
surgery; this finding indicates the existence in cured mice
of increased host's responses against small metastatic tumor
burden (61). This view is also supported by the fact the
antimetastatic effects of DM-COOK are significantly less
pronounced in mice immunosuppressed with cyclophosphamide
(63).

Toxicity

Some aspects of toxicity on the host have been examined
for the treatment with dimethyltriazenes used with schedules
and dosages effective for adjuvant antimetastatic treatment.
In mice subjected to leg amputation, the pre-operative
daily treatment with DTIC for 11 days significantly reduces
the number of erythrocytes, leukocytes and platelets, as
determined at the end of treatment; the reduction is still
significant on leukocytes and platelets 3 weeks after the
end of treatment. In these conditions, DM-COOK does not
cause any significant effect in comparison with drug untreated controls (5).
The toxicity on bone marrow, spleen and intestinal
mucosa has been examined by means of the measurement of the
fractional incorporation of ^{3}H-labeled thymidine. DTIC
causes a significant inhibition on bone marrow, which is not
observed after DM-COOK (7). Moreover, as observed histologically in mice, DM-COOK unlike DTIC does not cause significant hepatotoxicity, which in some instances was found to
severely restrict the clinical use of DTIC (64).
DTIC and DM-COOK have also been compared for their
immunodepressive properties, in terms of prolongation in
rejection time of allogeneic skin grafts in mice. The effects of DM-COOK are not statistically significant or significantly less pronounced than those of DTIC, depending on
the schedule and timing of treatment in relation to the time
of skin transplantation (65).
Finally, the mutagenic activity of DTIC and DM-COOK
have been determined in bacterial systems. On *S. typhimurium*
strains, the mutagenic activity of DM-COOK is moderate, is
significantly smaller than that of DTIC, and is only marginally increased by metabolic activation; similar findings
have been obtained using *E. coli* strains. For both bacterial
species, the concentrations required for the mutagenic
activity of DM-COOK are 2-20 times greater than those of
DTIC (66).

Conclusions

These results indicate that in mice bearing solid
malignant neoplasms the pre- and intra-operative treatment
with DTIC and DM-COOK is effective in causing an appreciable
amount of cures when combined with surgical removal of
primary tumor. The post-operative treatment is ineffective,
because of the mechanism of the antimetastatic action of

dimethyltriazenes which requires a direct interaction of the drug with the primary tumor. On the other hand, the pre-operative antimetastatic treatment is markedly effective also when started on clinically detectable tumors, and is still significant with largely advanced tumors. The antimetastatic effects of DM-COOK appear to require the cooperation with host responses left active after treatment. These effects are obtained at the expenses of a limited systemic toxicity, which for DM-COOK is very moderate and significantly less pronounced than that of DTIC.

GENERAL CONCLUSIONS AND DISCUSSION

A relatively large number of detailed investigations is available, showing that dimethyltriazenes cause selective antimetastatic effects in mice bearing solid malignant tumors. Dimethyltriazenes appear to have a broad spectrum of antimetastatic action, which is exerted upon several tumors which were examined. The antimetastatic action of dimethyltriazene is specific, and is dissociated from their cytotoxic effects on tumor cells; the structure-activity relationship analysis indicates optimal efficacy for hydrosoluble derivatives such as DM-COOK. The general validity of these findings is also indicated by the responsiveness of a panel of transplantable mouse leukemias to the antimetastatic effects of dimethyltriazenes. Indeed, dimethyltriazenes cause an increase in survival time which does not correlate with the cytotoxic reduction in the number of tumor cells, and which is associated with a dramatic reduction in the leukemic infiltration of crucial organs, such as the brain and liver (13,64,67,68).

In spite of the fact that dimethyltriazenes undergo significant *in vivo* (bio)transformation, their *in vivo* antimetastatic properties are neither attributable to hepatic metabolic conversion to monomethyltriazenes, nor to the extracellular generation of aryldiazonium cations. The occurrence, and the relevance, of (bio)transformation of dimethyltriazenes within tumor cells may be of importance, and has not been ascertained so far.

The pre- and intra-operative treatment with dimethyltriazenes combined with surgical removal of primary tumor results in a large proportion of cures for several animal-tumor systems examined. Natural antitumor host responses appear to participate in the antimetastatic effects of dimethyltriazenes, but the direct interaction of these drugs with tumor cells is required. The antimetastatic effects of DTIC and DM-COOK persist also after the cessation of drug administration, with a relatively stable loss of metastatic potential for tumor cells.

The mechanism by which these effects are exerted in tumor cells is not clarified so far. Genetic and/or epigenetic alterations of tumor cell properties are likely, and chemical xenogenization as described for mouse transplantable leukemias appears to be only marginally involved.

These data thus suggest that a specific antimetastatic action can be included in the antitumor clinical responses to DTIC. They also indicate that DM-COOK is an advantageous analog for DTIC in causing in mice selective antimetastatic effects at the expenses of reduced host toxicity. The antimetastatic action is persistent after cessation of treatment, and is significantly effective in combination with surgical removal of primary tumor when performed pre- and intra-operatively also on relatively advanced tumors. The post-operative treatment of already present (micro)metastasis is largely ineffective. These data provide an indication for the possible clinical use of dimethyltriazenes, and DM-COOK in particular, as antimetastatic adjuvants to be included into protocols of combined treatment of solid malignant tumors for the prophylaxis of tumor metastatic spread during the period between diagnosis and surgery. The post-operative treatment might be also considered, with the aim of reducing the metastatic potential of tumor cells in metastatic foci in order to prevent the formation of secondary metastasis. The indications and limitations of such an approach are not specific for dimethyltriazenes, but are evident for any drug treatment inhibiting tumor metastasis with a non-cytotoxic mechanism. The overall clinical implications and utility of such selective antimetastatic treatments have been accurately discussed by Hellmann (69). The effectiveness observed for ICRF 159 in a clinical trial designed for evaluating its antimetastatic effects in patients with colorectal carcinoma (70) is encouraging in suggesting to perform clinical trials aiming to evaluate the clinical applicability of the use of dimethyltriazenes as antimetastatic agents.

ACKNOWLEDGEMENTS. The work performed by the authors of this contribution has been performed under contract with the Italian National Research Council, Special Project Oncology (last contract n° 88.00691.44), and by grants from the Italian Ministry of Education (MPI 40 and 60%).

ABBREVIATIONS USED

BRL 51308: 4-carbethoxy-5-(3,3-dimethyl-1-triazeno)-2-methylimidazole
CCNU: 1-(2-chloroethyl)-3-cyclohexyl-1-nitrosourea
cisplatin: *cis*-dichlorodiammine platinum(II)
DM-CH$_3$: 1-*p*-tolyl-3,3-dimethyltriazene
DM-CONH$_2$: 1-*p*-carboxamidophenyl-3,3-dimethyltriazene
DM-COOK: 1-*p*-(3,3-dimethyl-1-triazeno)benzoic acid potassium salt
DM-NO$_2$: 1-*p*-nitrophenyl-3,3-dimethyltriazene
DTIC: 5-(3,3-dimethyl-1-triazeno)imidazole-4-carboxamide
GANU: 1-(β-D-glucopyranosyl-3-(2-chloroethyl)-3-nitrosourea
ICRF 159, Razoxane: (±)-1,2-di(3,5-dioxopiperazin-yl)propane
MM-CH$_3$: 1-*p*-tolyl-3-methyltriazene
MM-CONH$_2$: 1-*p*-carboxamidophenyl-3-methyltriazene
MM-COOK: 1-*p*-(3-methyl-1-triazeno)benzoic acid potassium salt
MM-NO$_2$: 1-*p*-nitrophenyl-3-methyltriazene

REFERENCES

1) Y.F. Shealy, J.A. Montgomery and W.R. Laster Jr., Antitumor activity of triazenoimidazoles, Biochem. Pharmacol. 11: 674 (1962).

2) J. Heyes, Antimetastatic effect of 4-carbethoxy-5(3,3-dimethyl-1-triazeno)-2-methylimidazole, J. Natl. Cancer Inst. 53: 279 (1974).

3) T. Giraldi, P. J. Houghton, D. M. Taylor and C. Nisi, Antimetastatic action of some triazene derivatives against the Lewis lung carcinoma in mice, Cancer Treat. Rep. 62: 721 (1978).

4) T. Giraldi, G. Sava, R. Cuman, C. Nisi and L. Lassiani, Selectivity of the antimetastatic and cytotoxic effects of 1-p-(3,3-dimethyl-1-triazeno)benzoic acid potassium salt, (+)-1,2-di(3,5-dioxopiperazin-1-yl)propane, and cyclophosphamide in mice bearing Lewis lung carcinoma, Cancer Res. 41: 2524 (1981).

5) G. Sava, T. Giraldi, L. Lassiani and C. Nisi, Antimeta-static action and hematological toxicity of p-(3,3-dimethyl-1-triazeno)benzoic acid potassium salt and 5-(3,3-dimethyl-1- triazeno)imidazole-4-carboxamide used as prophylactic adjuvants to surgical tumor removal in mice bearing B16 melanoma, Cancer Res. 44: 64 (1984).

6) T. Colombo, S. Garattini, L. Lassiani and M. D'Incalci, Activity of 1-p-(3,3-dimethyl-1-triazeno) benzoic acid potassium salt in M 5076/73A ovarian reticular cell sarcoma of the mouse, Cancer Treat. Rep. 66:1945 (1982).

7) T. Giraldi, G. Sava, L. Perissin, S. Zorzet, M. Tamaro and L. Dolzani, DTIC e suoi derivati: farmaci citotos-sici o antimetastatici?, Giornale Italiano di Chemioterapia 32: 43 (1985).

8) L. Lassiani, C. Nisi, T. Giraldi, G. Sava and R. Cuman, Selective antimetastatic triazenes: a quantitative approach, Quant. Struct.-Act. Relat. 3: 59 (1984).

9) G. Sava, T. Giraldi, L. Lassiani and R. Dogani, Effects of isomeric aryldimethyltriazenes on Lewis lung carcinoma growth and metastases in mice, Chem. Biol. Interactions 46: 131 (1983).

10) A. Gescher and M. D. Threadgill, The metabolism of triazene antitumor drugs, Pharmac. Ther. 32: 191 (1987).

11) G. F. Kolar and R.Z. Preussmann, Validity of a linear Hammet plot for the stability of some carcinogenic 1-aryl-3,3-dimethyltriazenes in aqueous system, Z. Naturforsch. 26b: 950 (1971).

12) T. Giraldi, A. M. Guarino, C. Nisi and G. Sava, Antitumor and antimetastatic effects of benzenoid triazenes in mice bearing Lewis lung carcinoma, Pharmacol. Res. Commun. 12: 1 (1980).

13) G. Sava, T. Giraldi, L. Lassiani and C. Nisi, Metabolism and mechanism of the antileukemic action of isomeric aryldimethyltriazenes, Cancer Treat. Rep. 66: 1751 (1982).

14) G. Sava, T. Giraldi, L. Lassiani and C. Nisi, Mechanism of the antimetastatic action of dimethyltriazenes, Cancer Treat. Rep. 63: 93 (1979).

15) B. L. Pool, Microsomal mediated metabolism of dialkyla-
     ryltriazenes. I. Demethylation of ring-halogenated
     3,3-dimethyltriazenes, J. Cancer Res. Clin. Oncol. 93:
     215 (1979).

16) B. L. Pool, Microsomal mediated metabolism of dialkyla-
     ryltriazenes. II. Isolation and identification of
     metabolites of 3,3-dimethyl-1-phenyltriazene, J. Cancer
     Res. Clin. Oncol. 93: 221 (1979).

17) T. Giraldi, C. Nisi and G. Sava, Investigation on the
     oxidative N-demethylation of aryl triazenes in vitro,
     Biochem. Pharmacol. 24: 1793 (1975).

18) G. Sava, S. Zorzet, L. Perissin, T. Giraldi, L. Lassia-
     ni, Effects of an inducer and an inhibitor of hepatic
     metabolism on the antitumor action of dimethyltria-
     zenes, Cancer Chemother. Pharmacol. 21: 241 (1988).

19) G. Abel, T. A. Connors and T. Giraldi, In vitro meta-
     bolic activation of 1-p-carboxamidophenyl-3,3-dimethyl
     triazene to cytotoxic products, Cancer Letters 3: 259
     (1977).

20) G. Sava, T. Giraldi, L. Lassiani, C. Nisi and P. B.
     Farmer, Mechanism of the antileukemic effects of 1-p-
     carboxamidophenyl-3,3-dimethyltriazene and its in vitro
     metabolites, Biochem. Pharmacol. 31: 3629 (1982).

21) M. B. Donati, A. Poggi, L. Mussoni, G. deGaetano and S.
     Garattini, Hemostasis and experimental cancer dissemi-
     nation, in: "Cancer invasion and metastasis: biological
     mechanism and therapy," S. B. Day, W. P. L. Myers, P.
     Stansly, S. Garattini and M. G. Lewis, eds., Raven
     Press, New York (1977).

22) M. B. Donati, J. F. Davidson and S. Garattini, "Malig-
     nancy and the hemostatic system," Raven Press, New York
     (1981).

23) P. Hilgard and R. D. Thornes, Anticoagulants in the
     treatment of cancer, Eur. J. Cancer 12: 755 (1976).

24) B. Maat and P. Hilgard, Anticoagualnts and experimental
     metastases; evaluation of antimetastatic effects in
     different model systems, J. Cancer Res. Clin. Oncol.
     101: 275 (1981).

25) T. Giraldi, G. Sava, E. Mitri and R. Cherubino, Hemos-
     tasis and mechanism of action of selective antime-
     tastatic drugs in mice bearing Lewis lung carcinoma,
     Eur. J. Cancer Clin. Oncol. 20: 961 (1984).

26) T. Giraldi, G. Sava, R. Cherubino, G. Bottiroli and G.
     Mazzini, Effects of DTIC, DM-COOK and ICRF-159 on the
     number of circulating Lewis lung carcinoma cells de-
     tected by flow cytometry, Clin. Expl. Metastasis 2: 151
     (1984).

27) B. S. Wang, G. A. McLoughlin, J. P. Richie and J. A.
     Mannick, Correlation of the production of plasminogen
     activator with tumor metastasis in B16 mouse melanoma
     cell lines, Cancer Res. 40: 288 (1980).

28) B. F. Sloane, J. R. Dunn and K. V. Honn, Lysosomal
     cathepsin B: correlation with metastatic potential,
     Science 212: 1151 (1981).

29) D. Tarin, B. J. Hoyt and D. J. Evans, Correlation of
     collagenase secretion with metastatic-colonization

potential in naturally occuring murine mammary tumours, Br. J. Cancer 46: 266 (1982).

30) B. U. Pauli, D. E. Schwartz, E. J. M. Thonar and K. E. Kuettner, Tumor invasion and host extracellular matrix, Cancer Metastasis Rev. 2: 129 (1983).

31) T. Giraldi and G. Sava, Malignancy and tumor proteinases: effects of proteinase inhibitors, in: "Proteinases and their inhibitors. Structure, function and applied aspects," V. Turk and Lj. Vitale, eds., Mladinska knjiga-Pergamon Press, Ljubljana-Oxford, (1981).

32) T. Giraldi, G. Sava, L. Perissin and S. Zorzet, Primary tumor growth and formation of spontaneous lung metastases in mice bearing Lewis lung carcinoma treated with proteinase inhibitors, Anticancer Res. 4: 221 (1984).

33) T. Giraldi, C. Nisi and G. Sava, Antimetastatic effects of N-diazoacetyl-glycine derivatives in C57BL mice, J. Natl. Cancer Inst. 58: 1129 (1977).

34) T. Giraldi, A. M. Guarino, C. Nisi and L. Baldini, Selective antimetastatic effects of N-diazoacethylglycine derivatives in mice, Europ. J. Cancer 15: 603 (1979).

35) T. Giraldi, G. Sava and C. Nisi, Mechanism of the antimetastatic action of N-diazoacetylglycinamide in mice bearing Lewis lung carcinoma, Europ. J. Cancer 16: 87 (1980).

36) A. J. Salsbury, K. Burrage and K. Hellmann, Histological analysis of the antimetastatic effect of (+)1,2-bis(3,5-dioxopiperazin-1-yl)propane, Cancer Res. 34: 843 (1974).

37) G. Sava and T. Giraldi, Antitumor effects of Ganu and other nitrosourea derivatives against transplantable leukemias and solid tumors in mice, Cancer Chemother. Pharmacol. 10: 167 (1983).

38) T. Giraldi, G. Sava, L. Lassiani e C. Nisi, Effetti di farmaci antitumorali sulla crescita e formazione di metastasi di tumori sperimentali, in: "Apporto della ricerca di base al controllo della crescita neoplastica, Vol. I" Idelson, Napoli (1981).

39) F. Bartoli-Klugmann, G. Decorti, L. Perissin, S. Zorzet, G. Sava and T. Giraldi, Effects of antineoplastic drugs on the proteolytic activity of murine metastasizing tumors, Chemioterapia II: 363 (1983).

40) T. Giraldi, G. Sava, L. Perissin and S. Zorzet, Proteinases and proteinase inhibition by cytotoxic and antimetastatic drugs in transplantable solid metastasizing tumors in mice, Anticancer Res. 5: 355 (1985).

41) J. F. DiStefano, G. Beck, B. Lane and S. Zucker, Role of tumor cell membrane-bound serine proteases in tumor-induced target cytolysis, Cancer Res. 42: 207 (1982).

42) F.S. Steven, B.S. Brown, T.P. Hulley, S. Itzahaki, T. Giraldi and L. Lassiani, Inhibition of cell-surface neutral protease of Ehrlich ascites tumour cells by potassium p-(3,3-dimethyl-1-triazeno)benzoate, Biochemistry International 11: 153 (1985).

43) T. Giraldi, G. Sava, L. Perissin, S. Zorzet and F.S.

Steven, Tumor cell metastasis and surface neutral proteinase: effects on antimetastatic and antitumor drugs, Invasion Metastasis 5: 336 (1985).

44) T. Giraldi, G. Sava, M. Kopitar, A. Suhar, V. Turk and A. Baici, Methodologic problems encountered in the assay of proteinases in Lewis lung carcinoma, a mouse metastasizing tumor, Tumori 68: 381 (1982).

45) V. Grill, F. Mallardi, S. Zorzet, L. Perissin, and T. Giraldi, Morphological analysis of metastatic potential and antimetastatic drug effects in mice bearing two lines of Lewis lung carcinoma, Clin. Expl. Metastasis 5: 233 (1987).

46) K. Hellmann, P. G. Salsbury, A. J. Burrage, A. W. Le Serve and S. E. James, Drug-induced inhibition of haematogenously spread metastases, in: "Chemotherapy of cancer dissemination and metastasis," S. Garattini and G. Franchi, eds., Raven Press, New York (1973).

47) G. H. Heppner, Tumor heterogeneity, Cancer Res. 44: 2259 (1984).

48) C. J. Honsik and L. Olsson, Phenotypic heterogeneity of malignant tumors as revealed by monoclonal antibodies, Invasion Metastasis 2: 249 (1982).

49) J. E. Talmadge, The selective nature of metastasis, Cancer Metastasis Rev. 2: 25 (1983).

50) A. Sacchi, A. Corsi, M. Caputo and G. Zupi, In vitro and in vivo selection of two Lewis lung carcinoma cell lines, Tumori 65: 657 (1979).

51) G. Zupi, A. Corsi, A. Sacchi, L. Lassiani and T. Giraldi, Effects of dimethyltriazenes on in vitro Lewis lung carcinoma tumor lines with different metastatic capacity, Invasion Metastasis 4: 179 (1984).

52) G. Sava, T. Giraldi, G. Zupi and A. Sacchi, Effects of antimetastatic dimethyltriazenes in mice bearing Lewis lung carcinoma lines with different metastatic potential, Invasion Metastasis 4: 171 (1984).

53) M. C. Fioretti, P. Fuschiotti, R. Bianchi, M. Allegrucci, U. Grohmann, L. Romani and P. Puccetti, in: "Advances in immunomodulation," B. Bizzini and E. Bonmassar, eds., Pythagora Press, Roma-Milan (1988).

54) T. Giraldi, G. Sava, S. Zorzet, L. Perissin, P. Piccini, S. Pacor and V. Rapozzi, Tumor metastatic potential following treatment with selective antimetastatic drugs, Clin. Expl. Metastasis 6 (Suppl. 1): 91 (1988).

55) S. Zorzet, L. Perissin, P. Piccini, V. Rapozzi, S. Pacor, G. Sava and T. Giraldi, Tumour metastatic potential after treatment with selective antimetastatic drugs, Pharmacol. Res. 21: 457 (1989).

56) S. Zorzet, L. Perissin, V. Rapozzi, S. Pacor, M. G. Rodani, G. Sava and T. Giraldi, Tumor metastatic potential following treatment with selective antimetastatic drugs, Farmaci e Terapia VI (Suppl. 4): 92 (1989).

57) M. Allegrucci, P. Fuschiotti, P. Puccetti, L. Romani and M. C. Fioretti, Changes in the tumorigenic and metastatic properties of murine melanoma cells treated with a triazene derivative, Clin. Expl. Metastasis 7: 329 (1989).

58) G. Sava, T. Giraldi, L. Lassiani and C. Nisi, Metabo-

lism and mechanism of the antileukemic action of isomeric aryldimethyltriazenes, <u>Cancer Treat. Rep.</u> 66: 1751 (1982).

59) G. Sava, T. Giraldi, C. Nisi and G. Bertoli, Prophylactic antimetastatic treatment with aryldimethyltriazenes as adjuvants to surgical tumor removal in mice bearing Lewis lung carcinoma, <u>Cancer Treat. Rep.</u> 66: 115 (1982).

60) T. Giraldi, G. Sava, L. Lassiani and C. Nisi, Selectivity of the mechanism of action of antimetastatic drugs, <u>in</u>: "Current Chemotherapy and Immunotherapy. Proceedings of the 12th International Congress of Chemotherapy. Florence, Italy 19-24 july 1981, Vol. II," P. Periti and G. Gialdroni Grassi, eds., The American Society for Microbiology, Washington DC U.S.A. (1982).

61) T. Giraldi, G. Sava, R. Cherubino, L. Lassiani, G. Bottiroli and G. Mazzini, Metastasis: mitostatic drugs, <u>in</u>: "The Control of Tumour Growth and its Biological Bases," W. Davis, C. Maltoni and St. Tanneberger, eds., Akademie-Verlag, Berlin (1983).

62) T. Giraldi, G. Sava and R. Cherubino, Selective antimetastatic action of p-(3,3-dimethyl-1-triazeno)benzoic acid potassium salt (DM-COOK), <u>in</u>: "Proceedings of the 13th International Congress of Chemotherapy, Tom 16," K. H. Spitzy and K. Karrer, eds., Verlag H. Egermann, Vienna (1983).

63) T. Giraldi, G. Sava, L. Perissin and S. Zorzet, Role of host responses in the drug treatment of metastasis, <u>in</u>: "Cancer Metastasis, Biological and Biochemical Mechanisms and Clinical Aspects," G. Prodi, L.A. Liotta, P.L. Lollini, S. Garbisa, S. Gorini and K. Hellmann, eds., Plenum Press, New York and London (1987).

64) G. Sava, T. Giraldi, L. Perissin, S. Zorzet, F. Mallardi and V. Grill, Infiltration of liver and brain by tumor cells in leukemic mice: prevention by dimethyltriazenes and cyclophosphamide, <u>Tumori</u> 70: 477 (1984).

65) I. Hrsak, Rudjer Boskovic Institute, Zagreb, Yugoslavia; personal communication.

66) M. Tamaro, L. Dolzani, C. Monti-Bragadin and G. Sava, Mutagenic activity of the dacarbazine analog p-(3,3-dimethyl-1-triazeno)benzoic acid potassium salt in bacterial cells, <u>Pharm. Res. Commun.</u> 18: 491 (1986).

67) G. Sava, T. Giraldi, F. Bartoli-Klugmann, G. Decorti and F. Mallardi, Effects of p-(3,3-dimethyl-1-triazeno)benzoic acid potassium salt on leukemic infiltration of brain and liver in mice bearing P388 leukemia, <u>Eur. J. Cancer Clin. Oncol.</u> 20: 287 (1984).

68) G. Sava, T. Giraldi, L. Perissin, S. Zorzet and G. Decorti, Effects of antimetastatic, antiinvasive and cytotoxic agents on the growth and spread of transplantable leukemias in mice, <u>Clin. Expl. Metastasis</u> 5: 27 (1987).

69) K. Hellmann, Antimetastatic drugs: laboratory to clinic, <u>Clin. Expl. Metastasis</u> 2: 1 (1984).

70) K. Hellmann, J. Gilbert, M. Evans, P. Cassell and R. Taylor, Effect of razoxane on metastases from colorectal cancer, <u>Clin. Expl. Metastasis</u> 5: 3 (1987).

EFFECTS OF TRIAZENES ON IMMUNE RESPONSES

Stefania D'Atri*, Lucio Tentori*, Maria Tricarico* and

Enzo Bonmassar**

* Institute of Experimental Medicine, CNR, Rome, Italy
**Department of Experimental Medicine and Biochemical
  Sciences, II University of Rome, Rome Italy

INTRODUCTION

One of the major goals of tumor immunology is the
development of experimental procedures capable of enhancing
responsiveness of the host to cancer.
Since the scarce immunogenicity of spontaneous neoplasias is
considered one of the most important factors in the failure
of the host to reject his own tumor (1), several attempts
have been made to modify the antigenic phenotype of cancer
cells (2-7). Along this line, studies performed in our
laboratory have shown that in vivo (8-10) or in vitro (11)
treatment of murine lymphomas with 5(3,3-dimethyl-1-
triazeno)-imidazole-4-carboxamide (DTIC) and other triazene
compounds (TZC) results in marked increases of tumor-
associated immunogenicity. TZC-treated tumor cells elicit
strong transplantation resistance in vivo (12), and T cell-
mediated immune responses in vitro (13) in syngeneic hosts.
This phenomenon has been defined "chemical xenogenization"
(CX, 14), and experimental evidence indicates that it relies
on the appearance of novel drug-mediated tumor antigens
(DMTA, 13), possibly due to somatic mutations (15-18).
The exploitation of CX in immunochemotherapy models is
however limited by the marked immunodepressive activity of
TZC, (19) which prevents an effective response in the tumor-
bearing mice against the TZC-modified potentially
immunogenic cancer cells. It follows that a restoration of
the immunocompentence of TZC-treated animals is required for
a successful utilization of CX in immunochemotherapy models.
Previous studies have shown that DTIC given 1 day before
antigen administration was capable of inhibiting humoral
antibody production (19-21), allograft (19-21), and graft
versus host T-dependent responses (19,21); generation of
cytotoxic T lymphocytes (CTL) against alloantigens in vitro
(21), and responsiveness of spleen cells to mitogens (19).
On the contrary delayed-type hypersensitivity was not
affected by pretreatment with DTIC before sensitization
(20,21). The DTIC-induced immunodepression has been

described to last up to 60 days (21,22), suggesting that this agent could affect a lymphoid cell population with extremely long turnover rate.

A good restoration of host-versus-tumor responsiveness in lymphoma-bearing mice treated with DTIC, can be obtained when selected lymphotoxic and cytoreductive agents are combined with syngeneic lymphoid cell transfer (23).

In the present paper we further investigate the effects of DTIC on various immune functions in the murine model to better elucidate the mechanism of the drug-induced immunodepression, and to determine an appropriate treatment for reverting the immune damage. Moreover, since CX could be of potential value in the immunochemotherapy of human neoplasias, preliminary studies were carried out on the immunomodulating activity of the in vitro active TZC 4(3-methyl-1-triazeno) benzoic acid, potassium salt (MTBA), on natural and antigen-dependent cytotoxic responses of human peripheral blood lymphocytes.

MATERIALS AND METHODS

Mice

Male C57BL/6 (H-2^b) and DBA/2 (H-2^d) mice, 2-4 months old were obtained from Charles River Breeding Laboratories (Calco, Como, Italy).

Mouse Tumors

The chemically induced leukemia L5178Y (H-2^d) of DBA/2 origin, the NK-susceptible cell line YAC-1 (H-2^a) of AJ origin and the IL-2 dependent CTLL line, were all maintained in vitro in RPMI 1640 (GIBCO, Paisley, Scotland, UK) supplemented with 10% fetal calf serum (FCS) (GIBCO), 2mM L-glutamine, 10mM HEPES and 5x10^{-5}M 2-Mercaptoethanol, hereafter referred to as Tissue Culture Medium (TCM).

Human Tumors

The B lymphoblastoid cell line BSM, the Burkitt lymphoma DAUDI and the erithroleukemia K562 were maintained in RPMI 1640 supplemented with 10% FCS and 2mM L-glutamine (hereafter referred to as Complete Medium, CM) and were subcultured three times a week.

Reagents

5-(3,3'-dimethyl-1-triazeno) imidazole-4-carboxamide (DTIC) was kindly supplied by Drug Synthesis and Chemistry Branch, Division of Cancer Treatment (National Cancer Institute, NIH, Bethesda, MD, USA). The drug was mixed with citric acid and mannitol in the proportion of 1:1:0.375 w/w, dissolved in distilled water and injected i.p. (200 mg/kg) immediately after preparation; 4-(3-methyl-1-triazeno) benzoic acid, potassium salt (MTBA), kindly supplied by Prof. L. Lassiani (Institute of Pharmaceutical Chemistry, University of Trieste, Italy), was dissolved in RPMI 1640; mitomycin-C was kindly supplied by Kyowa Italiana

Farmaceutici (Milano, Italy) and dissolved in distilled water; recombinant IL-2 (rIL-2) was kindly provided by Hoffmann La Roche (Nutley, NJ, USA).
Anti-mouse Lyt2 (anti-CD8) fluorescein and biotin conjugate, anti mouse L3T4 (anti-CD4) phycoerythrin (PE) conjugate, goat anti-mouse Ig fluorescein conjugate mAb, and streptavidin PE were purchased from Becton Dickinson (Mountain View, CA, USA). Sheep anti-rat IgG (H+L) affinity purified FITC conjugated was obtained from Sera Laboratories (Crawley Down, Sussex, U.K.). Anti-CD3 affinity purified fluorescein conjugate from 145-2C11 hybridoma (24), anti-IL-2R purified from 3C7 and 2E4 hybridomas (25), were a kind gift from Dr. Ada M. Kruisbeek (Biological Response Modifier Program, NCI, NIH, Bethesda MD, USA).

Splenocyte Preparation and In Vitro Generation of Mouse CTL

Spleens were removed from mice and splenocyte suspensions were prepared from pooled organs by mincing in RPMI 1640 containing 5% FCS. After filtration through nylon mesh, 1×10^7 responders (R) splenocytes were cocultered in 10 ml of TCM with mitomycin-treated ($25 \mu g/2 \times 10^6$ cells/ml) allogeneic splenocytes (stimulator, S) at R:S ratio of 1:1. Cocultures were incubated for 5 days at 37°C in a 5% CO_2 humidified atmosphere, and where mentioned in results rIL-2 (50U/ml) was added after two days of culture.

Proliferation Assay

Responder splenocytes (1×10^5) were cocultered in 0.2 ml of TCM with mitomycin-treated ($25 \mu g/2 \times 10^6$ cells/ml) allogeneic spleen cells at different 2-fold dilution concentrations from 1×10^5 to 0.25×10^5. Cultures were set up in U-shaped 96 well microtiter plates and incubated at 37°C in 5% CO_2 humidified atmosphere. After 48 hr 1 μCi of Methyl-^{3}H-deoxy-thymidine (^{3}H-TdR, specific activity 83 Ci/mmol, Amersham International Plc, Amersham, U.K.) was added to each well. The cells were harvested in glass fiber filter 12-16 hr later, and processed for standard liquid scintillation counting. The cell bound radioactivity was expressed as geometric mean counts per minute (cpm) of triplicate samples. Standard deviation (SD) never exceeded 10% of mean, and has been omitted.

Lymphokine Production Assays

IL-2 activity in supernatants from mixed lymphocytes cultures (MLC) described above was tested by measuring their effects on proliferation of CTLL cells. The supernatants to be tested (0.1 ml) were incubated with 3×10^3 CTLL cells (0.1 ml) in U-shaped 96 well microtiter plates. After 24 hr 1 μCi of ^{3}H-TdR was added to each well and radioactivity incorporation was determined as described above.

Immunofluorescence Staining and Flow Microcitofluorimetric (FMC) Analysis

Cells (1×10^6) were incubated for 30 min at 4°C with saturating amounts of fluorescein conjugated mAb. The cells were then washed twice, resuspended and analyzed for fluorescence. For two color analysis, cells were sequencially incubated with PE conjugated mAb, or with biotin conjugated mAb followed by streptavidin PE staining. Anti-human CD4 mAb FITC conjugated (Coulter lab. Hialey, FL, USA), and anti-human CD8 mAb PE conjugated, (Becton and Dickinson) were used as negative controls in single and double staining in every studied populations. These procedures were performed in PBS containing 0.1% bovine serum albumine and 0.02 % sodium azide.

FMC analysis was performed using an EPICS (Coulter Co., Hialeah, FL, USA) interfacied to a MDADS Coulter System. Data were collected on 20000 viable cells, as determined by forward light scatter and Propidium Iodide gating.

Preparation of Human MNC and In Vitro Treatment with MTBA

Peripheral blood mononuclear cells (MNC) were separated from buffy-coats obtained from healthy donors on a Ficoll-Hypaque gradient (26), washed twice and resuspended in CM at a concentration of 4×10^6 cells/ml. To MNC suspensions were then added equal volumes of RPMI 1640 alone or RPMI 1640 containing the appropriate amount of MTBA. The mixtures were incubated for 1 hr at 37°C in a Dubnoff metabolic shaker. At the end of the incubation period MNC were washed twice and tested for NK activity or for their ability to generate LAK cells and allo-CTL. Responsiveness to mitogens was also evaluated.

Generation of Human LAK Cells and Allo-CTL

For LAK generation control or MTBA-treated MNC were suspended (1×10^6 cells/ml) in RPMI 1640 containing 10% human AB serum and glutamine. MNC were incubated at 37°C for 72 hr with 100U/ml rIL-2, and then tested for cytotoxic activity against the NK-resistant and LAK-susceptible cell line Daudi. For CTL generation, MNC were cocultured in the same medium (2×10^6 cells/ml) with mitomycin-treated (50 μg/1x10^6 cells/ml) BSM cells for 5 days. The R:S ratio was 40:1. At the end of incubation period effector MNC were recovered and tested for cytotoxicity against fresh BSM cells.

Responsiveness to Mitogens

MNC (1×10^5) suspended in 0.2 ml of CM were incubated at 37°C in U-shaped 96 well microtiter plates in the presence of 10 μg/ml PHA. After 48 hr 1 μCi/well of ^{3}H-TdR was added, and 18 hr later radioactivity incorporation was evaluated as mentioned above.

Cytotoxicity Assay

Target cells were removed from continuous culture, labeled by incubation with 100 μCi of $Na_2^{51}CrO_4$ (Amersham) in a 5% CO_2 humidified atmosphere for 1 hr at 37°C. After incubation, the cells were washed three times and resuspended in CM at a final concentration of 2×10^4 cells/ml. Effector cells in 0.1 ml of CM were plated in quadruplicate in U-shaped 96 well microtiter plates by making serial two-fold dilution starting at a concentration of 2×10^6 cells/ml. Labeled target cells were then added in a volume of 0.1 ml to give a final volume of 0.2 ml and an effector (E) to target (T) cell ratio ranging from 100:1 to 12.5:1. The plates were then centrifuged at 80 x g for 5 min, and incubated for 4 hr at 37°C in 5% CO_2 humidified atmosphere. Then the plates were centrifuged at 250 x g and 0.1 ml of supernatant was counted in a gamma scintillation counter. The means of quadruplicate samples were expressed as the percentage of specific lysis according to the formula:

$$\% \text{ specific lysis} = \frac{\text{test cpm} - \text{autologous cpm}}{\text{total cpm}} \times 100$$

Where test cpm is the mean cpm released in presence of effector cells; autologous cpm is the mean cpm released by target cells incubated with 2×10^5 unlabeled autologous cells in place of effector cells and total cpm is the total amount of ^{51}Cr incorporated in target cells.

Calculation of Killed Cells (KC)

Dose-response curves were obtained by plotting the percentage of specific lysis (n%) vs ln x, where x is the number of effector cells per well. The number of target cells killed by a fixed number (m) of effector cells, i.e. KC(m) was calculated as follows:

$$KC(m) = \frac{m \times T \times n\%}{En \times 100}$$

where T is the total number of target cells present in each well and En is the number of effector cells present in each well, at the selected E:T ratio, n% is the specific lysis produced by En effector cell extrapolated from the dose-response curve. In the present study the E:T ratio used for KC(m) calculation was 100:1.
Covariance analysis of regression lines based on dose-response curves was used to perform statistical analysis on KC(m) values.

RESULTS

Effect of In Vivo DTIC Treatment on the Generation of CTL Against Alloantigens In Vitro

Male C57Bl/6 ($H-2^b$) mice were treated with a single ip. injection of DTIC (200 mg/kg). Spleens were collected 1 day after drug administration and tested for the generation of CTL following 5 days of coculture in vitro with inactivated allogeneic DBA/2 ($H-2^d$) spleen cells. At the end of coculture, effector cells were tested for cytotoxicity against the tumor line L5178Y of DBA/2 origin, the NK-susceptible YAC-1 cells, or against Con A stimulated syngeneic spleen cells, as negative control.

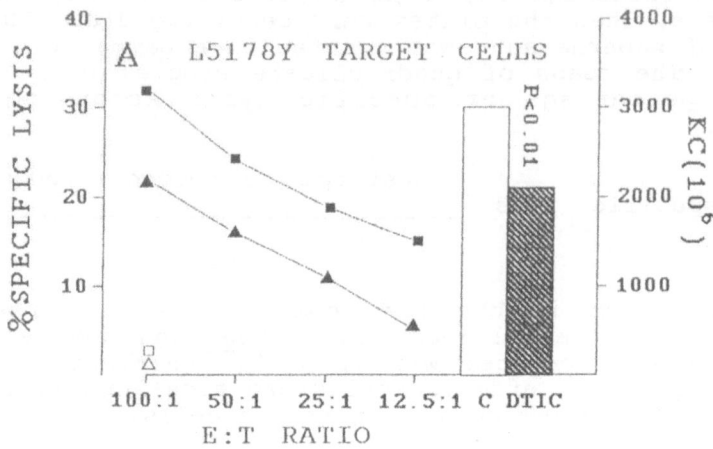

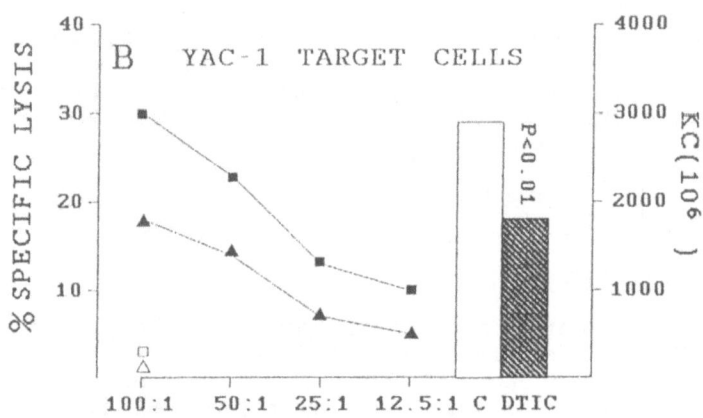

Figure 1. Effect of in vivo DTIC-treatment on CTL generation against alloantigen in vitro. Splenocytes from C57Bl/6 (H-2^b), control (■) or DTIC-treated mice (▲), were cultured in vitro with mitomycin-treated splenocytes from DBA/2 (H-2^d) mice for 5 days. Effector (E) cells were then assayed against L5178Y (Fig. 1A) or YAC-1 (Fig. 1B) ^{51}Cr-labelled target (T) cells. Data are expressed in terms of percentage of specific target cell lysis at different E:T ratios, and in terms of target cells killed by 1x10^6 effectors at the E:T ratio of 100:1, i.e. KC(10^6). The percentage of specific lysis of control (□) or DTIC-treated (△) non-immune splenocytes at the E:T ratio 100:1 is also reported.

The results of a representative experiment are illustrated in Figure 1. The data, expressed in terms of percentage of specific lysis at different E:T ratios and in terms of KC(10^6) at the E:T ratio of 100:1, confirm the depressive effects of DTIC treatment on the generation of allo-CTL. Strong inhibition of CTL generation against DBA/2 allo-antigens (Figure 1A), and of non-MHC restricted cytotoxicity against YAC-1 target (Figure 1B) is observed in DTIC-treated mice. Cytolysis of control group against syngeneic ConA-stimulated spleen cells never exceeded 10% (data not shown).

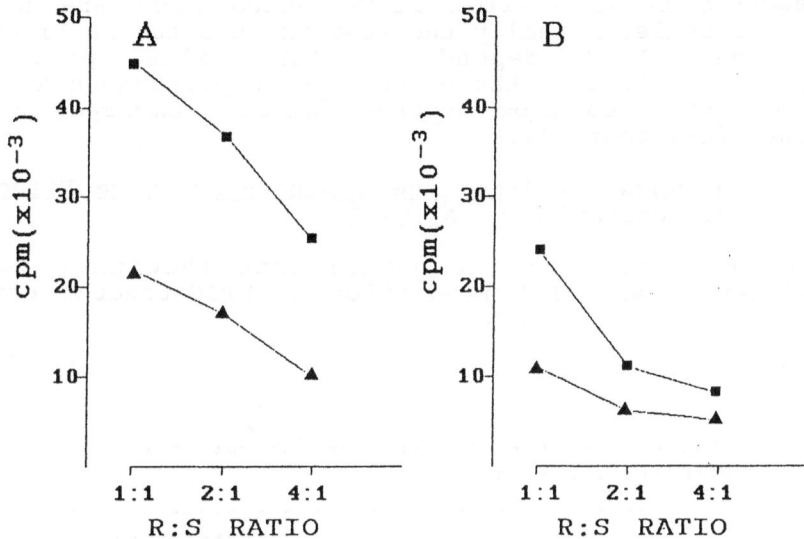

Figure 2. Effects of in vivo DTIC-treatment on proliferation and IL-2 production by spleen cells. Figure 1A: proliferation of 1 x 10^5 splenocytes from control and DTIC-treated C57Bl/6 mice, cocultured with graded numbers of DBA/2 allogeneic mitomycin-treated stimulator splenocytes. ^{3}H-TdR incorporation was determined after 48 hr. Each value (■, controls; ▲, splenocytes collected one day after DTIC treatment) represents the geometric mean of triplicate cultures; the SE were always less than 10%. Means cpm of responder splenocytes incubated with medium alone were 3331 and 3224 for control and DTIC-treated group respectively. Figure 2B: supernatants from MLC (see above) were tested for the ability to support proliferation of 1 x 10^3 CTLL cells. Data are expressed as geometric means of ^{3}H-TdR incorporation of triplicate cultures determined after 24 hr. The SE were always less than 10%. Positive controls consisted in proliferation of 1 x 10^3 CTLL cells in the presence of 3U/ml of rIL-2 (mean 4.47 x 10^4 cpm). ^{3}H-TdR incorporation by CTLL incubated in the presence of supernatants collected from non-immune groups was comparable to background levels.

Effect of In Vivo DTIC Treatment on Lymphocyte Proliferation and IL-2 Production in Response to Alloantigens In Vitro

C57Bl/6 spleen cells obtained from control or DTIC-treated mice (200 mg/kg) 1 day after drug administration, were cocultivated with inactivated DBA/2 spleen cells for 72 hr in 96-well plates. During the last 18 hr, ^{3}H-TdR was added to evaluate effector cell proliferation. IL-2 activity in the supernatants collected from the same cultures after 48 hr was also tested. Figure 2A shows that DTIC strongly inhibits the proliferation of C57Bl/6 splenocytes in response to graded number of stimulator cells. In the same cocultures, IL-2 production appears to be reduced when splenocytes obtained from DTIC-treated mice are used as responder cells. Actually the results reported in Figure 2B, show that IL-2 dependent CTLL cells proliferate significantly less in the presence of supernatants harvested from MLC obtained from treated animals compared to those obtained from controls.

IL-2 Can Reverse In Vitro the Immunosuppressive Effects of DTIC on the Generation of Allo-CTL

On the basis of the experiments showing a reduced proliferation and IL-2 production by DTIC-treated effector

Table 1. IL-2 can reverse in vitro the immunosoppressive effects of DTIC treatment in vivo

Mice[a]	Sensitizer	IL-2[b]	Cytotoxicity	
			L5178Y	YAC-1
Control	-	-	1139 (2354- 552)	<10
	-	50U/ml	1629 (3565- 745)	1453 (1583-1328)
	DBA/2	-	5206 (5502-4926)	3057 (3496-2674)
	DBA/2	50U/ml	6043 (7052-5178)	3953 (4887-3197)
DTIC/treated	-	-	1340 (1424-1260)	<10
	-	50U/ml	2340 (2605-2101)	<10
	DBA/2	-	3553 (4277-2951)	1207 (1402-1040)
	DBA/2	50U/ml	5109 (6086-4290)	3109 (3917-2468)

[a] Spleen cells from intact or DTIC-treated C57Bl/6 (H-2^b) mice were cocultured for 5 days with medium alone or with DBA/2 (H-2^d) mitomycin-treated splenocytes (R:S ratio 1:1) and then tested for cytotoxic activity against L5178Y or YAC-1 target cells.
[b] Recombinant IL-2 (50U/ml) was added after 48 hr from the onset of cocultures.
[c] Data are expressed in terms of KC(10^6). In parenthesis: Mean KC(10^6) plus Standard Deviation (SD), Mean KC(10^6) minus SD.

cells in response to allostimuli, and taking into account that IL-2 appears to be necessary to promote growth and acquisition of cytolytic activity of antigen specific CTL (27), we investigated whether the addition of rIL-2 during the generation of allogeneic CTL in vitro, could reverse the immunodepressive effects of DTIC treatment of the host. The results reported in Table 1 show that the addition of 50U/ml of rIL-2 48 hr after the onset of the allogeneic coculture leads to an increase in the cytolytic activity of CTL against the specific L5178Y target cells both in the control and DTIC-treated group. Moreover, cytotoxicity of CTL generated in the presence of rIL-2 using effectors from DTIC-treated hosts is comparable to that of CTL generated in the absence of rIL-2 using effector from control mice. Non immune responder cells showed a very low capacity to lyse L5178Y target cells also when rIL-2 were added to the culture.
Similar results were obtained using the NK-susceptible YAC-1 target cells (Table 1).

Table 2. FMC analysis on spleen cell subpopulations from MLC performed with untreated or DTIC-treated responder splenocytes[a].

	Percentage of positive cells[b]			
Cell phenotype	Control	DTIC	IL-2	DTIC + IL-2
CD4+	32.5	44.1	47.1	55.2
CD8+	50.3	36.2	46.4	35.1
IL-2R+	18.8	18.2	48.1	41.7
IL-2R+CD4+	9.3	13.6	18.7	20.2
IL-2R+CD8+	9.5	4.6	30.0	21.5
IL-2R-CD4+	28.8	30.5	29.0	35.0
IL-2R-CD8+	40.8	31.6	14.4	13.6

[a] Splenocytes from control or from DTIC-treated C57Bl/6 mice were cocultured for 5 days with allogenic DBA/2 spleen cells. Where indicated, rIL-2 (50U/ml) was added to MLC at day 2.

[b] For two-color FMC analysis, cells were sequentially incubated with a combination of rat anti-IL-2R (3E4 + 3C7) mAb purified from ascites, followed by FITC-sheep anti-rat (H + L) and then by anti-CD4, PE conjugated or by anti-CD8 biotin conjugated plus streptavidin PE. The percentage of IL-2R positive cells from responder splenocytes cultured in absence of stimulator cells was 10.4 in the culture from control and 9.6 in the culture from DTIC-treated mice. When rIL-2 was added after 48 hr of the culture the percentage of IL-2R positive cells was 13.7 and 10.5 respectively.

IL-2R Expression in Effector Splenocytes from Control or DTIC-treated Mice after the 5-day Allo-MLC .

To better elucidate the effect of DTIC treatment on the IL-2/IL-2R patway, splenocytes cocultured for 5 days with allogeneic cells were analyzed for their expression of CD4, CD8 and IL-2R surface markers. Table 2 shows that higher CD4+ and lower CD8+ percentages with respect to controls were found in allo-MLC performed with splenocytes form DTIC-treated mice. Similar results were obtained for MLC carried out in the presence of rIL-2.
IL-2R expression is not different when control and DTIC-treated groups are compared at the level of bulk population. However, the CD8+ IL-2R+ subset is reduced in the drug-treated splenocytes. The addition of rIL-2 to the coculture significantly increases the percentages of IL-2R+ cells and of CD8+ IL2R+ subset in both control and treated groups. In this case even though the percentages of positive cells are lower in MLC performed with treated splenocytes with respect to control MLC plus rIL-2, they are higher than those found in the control MLC not exposed to rIL-2.

Table 3. Effects of MTBA on human natural and antigen-dependent cell-mediated cytotoxicity.

Immune function	rIL-2	Sensitizer	Target cells	Cytotoxicity [d] Control MNC	P_1 [e]	MTBA-treated MNC [f]	P_2 [e]
NK activity [a]	-	-	K562	5302	-	4495	<0.01
LAK generation [b]	-	-	DAUDI	1096	-	410	<0.01
	100U/ml	-	DAUDI	7025	<0.01	2800	<0.01
CTL generation [c]	-	-	BSM	1444	-	390	<0.01
	-	BSM	BSM	6431	<0.01	2638	<0.01

[a] NK activity of MNC was tested immediately after MTBA treatment.
[b] Control or MTBA-treated MNC were incubated for 72 hr in the presence of 100U/ml rIL-2 or in medium alone. Cytotoxicity was then tested against the NK-resistant and LAK-susceptible cell line DAUDI.
[c] Control or MTBA-treated MNC were incubated for 5 days with mitomycin treated allogeneic BSM cells , or with medium alone. Effector cells were then assayed for cytotoxic activity against fresh BSM cells.
[d] Cytotoxic activity of effector MNC was tested using a 4 hr ^{51}Cr-release assay. Data are expressed in terms of target cells killed by 10^6MNC at the E:T ratio of 100:1, i.e. KC(10^6).
[e] Pn, P values calculated according to covariance analysis; P_1, activated MNC (LAK cells or CTL) vs unstimulated MNC; P_2, MTBA treated MNC vs control MNC.
[f] MNC were treated with 100 µg/ml MTBA for 1 hr at 37°C.

Table 4. Effect of MTBA on human MNC responsiveness to mitogens.

| MTBA µg/ml[a] | ^{3}H-TdR incorporation (cpm) | |
	Medium alone	PHA 10 µg/ml
-	2446	121579
50	2261	130822
100	1097	62027
200	222	2479

[a] MNC were treated with MTBA for 1 hr at 37°C, washed and incubated for 72 hr with PHA in 96 well-plates (2 x 10^{5} cells/well). During the last 18 hr 1 µCi/well ^{3}H-TdR was added. Each value represents the geometric mean of triplicate cultures. SE was always less than 10%.

Human Studies

To investigate the immunosuppressive effects of TZC on human natural and antigen-dependent cell-mediated cytotoxicity, peripheral blood MNC were incubated for 1 hr at 37°C in the presence of medium alone or of 100 µg/ml MTBA. Effector MNC were then washed and tested immediately for their NK activity against ^{51}Cr-labeled K562 target cells. Control or MTBA-treated MNC were also assayed for their ability to generate LAK cells in response to rIL-2 (100 U/ml, 72 hr), or to generate allo-CTL after a 5-day coculture with the lymphoblastoid cell line BSM. The results of a representative experiment are illustrated in Table 3. The data, expressed in terms of KC(10^{6}), show that MTBA strongly depresses NK activity of human MNC and inhibits the generation of LAK cells and CTL in response to the appropriate stimuli.
The effect of MTBA on MNC responsiveness to mitogens was also evaluated. The results of a typical experiment illustrated in Table 3, show that proliferation of MTBA treated MNC in response to PHA (10 µg/ml) is significantly reduced after MTBA treatment, in a dose-response manner.

DISCUSSION

The ultimate goal of experimental studies on TZC-induced CX is its application to cancer immunotherapy.
However a major limitation to the exploitation of CX in animal models resides in the severe immunodepression that follows in vivo treatment of the host with TZC. On these bases it appears of primary interest to investigate the mechanism of TZC-induced immunodepression with the aim to determine procedures capable of restoring the immunocompetence of the host.
Therefore in the present study the effects of DTIC on T cell phenotype, proliferation and lymphokine production in response to allo-stimuli were investigated in a murine

model, since so far no data were available on these parameters. Moreover we extended the studies to human models in vitro.

The results illustrated in this paper confirm the previously reported failure to generate CTL in response to allo-stimuli by responder cells from DTIC-treated mice (21). Furthermore the data show that DTIC exerts inhibitory effects on splenocyte proliferation and IL-2 production in MLC. Taking into account the crucial role played by IL-2 in T-T cell interaction (28) it is possible to hypothesize that inhibition of CTL generation could be due, at least in part, to the impairment of IL-2 dependent processes.

We therefore carried out experiments to investigate whether the addition of low amounts of rIL-2 could restore allo-CTL responses. Actually the addition of 50U/ml of rIL-2 to allo-MLC set up with responder splenocytes collected from DTIC-treated mice, results in cytolytic activity comparable to that of untreated controls not exposed to rIL-2. The presence of IL-2 enhances both specific and non specific (i.e. against YAC-1 target cells) cytotoxicity of effector cells obtained from control or DTIC-treated mice.

The possibility that DTIC-induced immunodepression would be caused also by impairment of IL-2R expression on responder cells has been considered. Data in Table 2 show that the total number of IL-2R+ cells is not affected in the treated group. However when IL-2R expression is assayed within the CD4+ and CD8+ subset, a decrease in the CD8+ IL-2R+ subset is observed. When rIL-2 is added to the culture, marked increase of IL-2R+ cells occurs in all groups. In particular the percentage of CD8+ IL2R+ subset of treated group appear higher than that of the control cultures not exposed to IL-2.

Taken together data suggest that DTIC could impair T cell responses through inhibition of IL-2 production/release and proliferation of CD8+ IL-2R+ cell subset, which presumably represents the majority of allo-CTL (28). The mechanism by which IL-2 can reverse the inhibitory effect of DTIC needs further investigations. It is possible that exogenous IL-2 can increase the lytic potential of CTL, or allow the expansion of CTL clones which need higher level of IL-2 for the acquisition of lytic activity.

The studies carried out on human model, show that peripheral blood MNC treated in vitro with MTBA, are profoundly inhibited in their NK activity, ability to generate LAK cells or allo-CTL and to proliferate in response to polyclonal activation by PHA. MNC obtained from different donors show variable sensitivity to MTBA, mainly at low concentrations of the drug (data not shown). It is possible that the activity of O^6-Alkylguanine-DNA alkyltransferase (29) could be related to this different susceptibility of MNC to MTBA. The findings that NK activity of MNC is also depressed by drug treatment suggest that MTBA can directely affect the lytic machinery of effector cells. However in the case of LAK and CTL generation, the inhibition of proliferation, and consequently of clonal expansion of cytotoxic cells, could represent the main mechanism of immunodepression. Studies are in progress to investigate whether lymphokine production, and proliferation of MNC in MLC is affected by MTBA, and whether IL-2 could restore MTBA inhibition of CTL generation.

In conclusion the data presented here show that TZC possess strong immunodepressive activity both in mouse and in man, and suggest that IL-2 could be efficiently used to restore host immunoresponsiveness. This would provide rational bases to immunochemotherapy protocols in which CX could be combined with IL-2 in order to attain immunological recovery of TZC-treated cancer bearing hosts.

ACKNOWLEDGMENT

This work was supported by a grant from Istituto Superiore di Sanita', Roma, Italy (Italia-USA Program on Therapy of Neoplasias). The authors wish to thank Mrs. Barbara Bulgarini for her expert assistance with the preparation of this manuscript.

REFERENCES

1. N. Galili, B. Devens, D. Noor, S. Becker, and E. Klein, Immune responses to weakly immunogenic virally-induced tumors. I Overcoming low responsiveness by priming mice with syngeneic in vitro tumor line or allogeneic cross-reactive tumor, Eur. J. Immunol. 8:17 (1978).
2. J. Lindeman, and P. Klein, Viral Oncolysis: Increase on immunogenicity of host cell antigen associated with influenza virus, J. Exp. Med. 126:93 (1963).
3. H. Kobayashi, T. Kodama, T. Shirai, H. Kaji, M. Hosokawa, F. Sendo, H. Saito, and N. Takeichi, Artificial regression of rat tumors injected with Friend virus (xenogenization): an effect produced by acquired antigen, Hokkaido J. Med. Sci. 44:133 (1969).
4. E. Benjamini, S. Fong, C. Erickson, C.Y. Leung, D. Rennick, R.J. Scibiensky, Immunity to lymphoid tumors induced in syngeneic mice by immunization with mitomycin C-treated cells, J. Immunol. 118:685 (1977).
5. T. Boon, and O. Kellermann, Rejection by syngeneic mice of cell variants obtained by mutagenesis of a malignant teratocarcinoma cell line, Proc. Natl. Acad. Sci. USA 74:272 (1977).
6. T. Boon, Antigenic tumor cell variants obtained with mutagens, Adv. Cancer Res. 39:121 (1983).
7. G.J. Bekesi, J.F. Holland, J.P. Roboz, Immunotherapy trials with neuraminidase modified tumor cells, in: "Immunological approaches to cancer therapeutics", E. Mihich ed., John Wiley, New York (1982).
8. E. Bonmassar, A. Bonmassar, S. Vadlamudi, and A. Goldin, Immunological alteration of leukemic cells in vivo after treatment with an antitumor drug, Proc. Nat. Acad. Sci. USA 66:1089 (1970).
9. E. Bonmassar, A. Bonmassar, S. Vladmudi, and A. Goldin, Antigenic changes of L1210 leukemia in mice treated with 5-(3,3'-Dimethyl-1-Triazeno)-Imidazole-4-Carboxamide, Cancer Res. 32:1446 (1972).
10. A. Bonmassar, L. Frati, M.C. Fioretti, L. Romani, A. Giampietri, and A. Goldin, Changes of immunogenic properties of K36 lymphoma treated in vivo with 5-(3,3'-Dimethyl-1-Triazeno)-Imidazole-4-Carboxamide (DTIC), Eur. J. Cancer 15:933 (1979).

11. B. Nardelli, A. R. Contessa, L. Romani, G. Sava, C. Nisi, and M.C. Fioretti, Immunogenic changes of murine lymphoma cells following in vitro treatment with aryl-triazene derivatives, Cancer Immunol. Immunother. 16:157 (1984).

12. C. Riccardi, M. C. Fioretti, A. Giampietri, A. Puccetti, and A. Goldin, Growth rejection patterns of murine lymphoma cells antigenically altered following drug treatment in vivo, Transplantation 25:63 (1978).

13. L. Romani, M.C. Fioretti, E. Bonmassar, In vitro generation of primary cytotoxic lymphocytes against L5178Y leukemia antigenically altered by 5-(3,3'-dimethyl-1-triazeno)imidazole-4-carboxamide, Transplantation 28:21 (1979).

14. P. Puccetti, L. Romani, M. C. Fioretti, Chemical xenogenization of tumor cells, Trends Pharmacol. Sci. 6:485 (1985).

15. A. Giampietri, M. C. Fioretti, A. Goldin, and E. Bonmassar, Drug-mediated antigenic changes in murine leukemic cells: antagonistic effect of quinacrine, an antimutagenic compound, J. Natl. Cancer Inst. 64:297 (1980).

16. M. C. Fioretti, R. Bianchi, L. Romani, and E. Bonmassar, Drug-induced immunogenic changes of murine leukemia cells: dissociation of onset of resistance and emergence of novel immunogenicity, J. Natl. Cancer Inst. 71:1247 (1983).

17. D. D. Beal, J. L. Skibba, W. A. Croft, S. M. Cohen, and G. T. Bryan, Carcinogenicity of the antineoplastic agent 5(3-3-dimethyl-1-triazeno)imidazole-4-carboxamide, and its metabolites in rats, J. Natl. Cancer Inst. 54:951 (1975).

18. B. H. Venger, C. Hansch, G. J. Hatheway, and Y. U. Amrein, Ames test of 1(X-Phenyl)3-3-dialkyltriazenes. A quantitative structure-activity study, J. Med. Chem. 22:473 (1979).

19. B. Nardelli, P. Puccetti, L. Romani, G. Sava, E. Bonmassar, and M. C. Fioretti, Chemical xenogenization of murine lymphoma cells with triazene derivatives: immunotoxicological studies, Cancer Immunol. Immunother. 17:213 (1984).

20. M. C. Fioretti, Immunopharmacology of 5-(3,3-dimethyl-1-triazeno)-imidazole-4-carboxamide (DTIC), Pharmacol. Res. Comm. 7:481 (1975).

21. A. Giampietri, and E. Bonmassar, Drug induced modulation of immune responses in mice, effects of 5-(3,3-dimethyl-1-triazeno)-imidazole-4-carboxamide (DTIC) and cyclophosphamide (CY), J. Immunopharmacol. 1:61 (1978).

22. P. A. Puccetti, and M. C. Giampietri, Long term depression of two primary immune responses induce by a single dose of DTIC, Experientia 34:799 (1978).

23. A. Giampietri, A. Bonmassar, P. Puccetti, A. Circolo, A. Goldin, and E. Bonmassar, Drug-mediated increase of tumor immunogenicity in vivo for a new approach to experimental cancer immunotherapy, Cancer Res. 41:681 (1981).

24. O. Leo, D.H. Foo, L.E. Sacks, and J.A. Bluestone, Identification of a monoclonal antibody specific for a murine T3 polipeptide, Proc. Natl. Acad. Sci. 84:1374 (1987).

25. G. R. Ortega, R.J. Robb, E.M. Shevach, and T.R. Malek, The murine IL-2 receptor. I. Monoclonal antibodies that define distinct functional epitopes on activated T cell and react with activated B cells, J. Immunol. 133:1970 (1984).
26. A. Boyum A., Isolation of lymphocytes from human blood. Further observation. Methylcellulose, dextran and ficoll as erythrocyte aggregating agents, Scand. J. Clin. Invest. 21:31 (1968).
27. H. Wagner, C. Heeg, and Hardt C., Multiple signals required in cytolytic T cell responses, in: "Progress in Immunology VI", L.Brent and J.Harborow eds., Academic Press Inc., (1986).
28. H. C. Wagner, C. Hardt, K. Heeg, K. Pfiezenmaier, W. Solbach, R. Bartlett, H. Stockinger, and M. Rollinghoff, T-T cell interactions during cytotoxic T lymphocyte (CTL) responses: T cell derived helper factor (Interleukin-2) as a probe to analyze CTL responsiveness and thymic maturation of CTL progenitors, Immunol. Rev. 51:215 (1980).
29. E. A. Valdstein, C. E. H. Cao, M. A. Bender, and R. R. Setlow, Abilities of extracts of human lymphocytes to remove O^6-methyl guanine from DNA, Mut. Res. 95:405 (1982).

XENOGENIZATION OF EXPERIMENTAL TUMORS BY TRIAZENE DERIVATIVES

Paolo Puccetti, Luigina Romani, Ursula Grohmann, Roberta
Bianchi, Patrizia Fuschiotti, Massimo Allegrucci, and
Maria Cristina Fioretti

Dept. Exp. Med. Bioch. Sci, Sect. Pharmacology, Univ. of
Perugia, Via del Giochetto, 06100 Perugia, Italy

INTRODUCTION

The antigenic phenotype of experimental tumors can be modified through
a variety of procedures that either directly - and rather transiently, as a
rule - affect the membrane structures of the cell or involve stable, often
hereditary, changes in the cell biology (e.g., the genetic code). The term
of chemical xenogenization (from the Greek xenos, meaning strange, foreign)
was first introduced by our group to indicate the appearance of stable
tumor-associated transplantation antigens on murine tumors subjected to
chemical treatment and thus rendered antigenically foreign to the host of
origin (1). In 1970 Bonmassar et al. (2) had indeed found that murine leukemia cells, on repeated in vivo exposure to the antitumor agent dacarbazine,
would become increasingly immunogenic such that they eventually acquired a
degree of foreignness capable of conditioning the histocompatible host to
reject challenge with the drug-treated tumor.

In this context, chemical xenogenization will indicate the induction
of stable tumor variants with increased immunogenicity following exposure
of the original (parental) neoplasm to different chemicals.

We will review evidence that chemical xenogenization may represent a
suitable means to increase the immunogenicity of tumor cells, and that this
phenomenon can be successfully exploited in experimental models of tumor
immunotherapy. It will be shown that many drugs or chemicals can indeed enhance the immunogenic strength of experimental tumors, most of which carry
new transplantation antigens. We will also discuss the possible application
of chemical xenogenization to the treatment of human cancer.

CHEMICAL XENOGENIZATION BY TRIAZENE DERIVATIVES

1. Dacarbazine

Many triazene derivatives, with either an imidazole or aryl moiety,
have been synthesized which possess both cytoreductive and xenogenizing
properties (3). The triazenyl-imidazole derivative DTIC (or dacarbazine)

was the first compound of this class to manifest a strong ability to induce immunogenic changes in murine lymphoma cells. The experimental model most suitable to reveal xenogenization by DTIC in vivo is the one originally described by Bonmassar et al. (2). In this model, treatment of tumor cells with DTIC leads to a progressive increase in the immunogenicity of the tumor so that previously nonimmunogenic inocula become capable of evoking a strong antitumor response which creates a state of specific resistance. In DTIC-treated animals, however, lethal tumor growth occurs since the highly immuno-depressive activity of the drug prevents host antitumor responses without affecting the growth of the xenogenized line that has become resistant to the cytoreductive activity of DTIC. Therefore, the newly acquired immunogenic potential of the xenogenized tumor is only revealed by grafting the drug-treated tumor into naive, immunocompetent hosts (4-10).

2. Aryl-triazenes

Studies concerning the relationship between chemical structure, cyto-reductive and xenogenizing properties of triazene compounds led to the synthesis of a class of derivatives in which the imidazole ring present in DTIC was replaced by an aryl moiety (11-13). Most of these compounds proved to be strong xenogenizing agents, thus suggesting that the imidazole ring is not mandatory for the activity (13). Subsequent studies also showed that dimethyltriazenes, including DTIC, are not active per se in vivo but require metabolic activation which is carried out by liver microsomal enzymes and leads to generation of monomethyl species. When a series of dimethyl aryl-triazenes and related monomethyl compounds were assayed for induction of xenogenization in vitro, it was found that the dimethyl derivatives required the presence of mouse liver microsomes whereas the corresponding monomethyl compounds did not (13). Similar results had been obtained with DTIC and its related monomethyl derivative (13). A number of data are now available which point to the strict analogy between the activity of DTIC and that of aryl-triazenes (Table 1), and also indicate that similar xenogenizing properties are shown by antitumor drugs and mutagenic chemicals belonging to different classes (14,15).

Table 1. Properties of triazene derivatives*

	Cytoreductive activity	Suppression of immunity:		Xenogenizing activity
		humoral	cell-mediated	
DTIC	+	++	+++	++
MIC	+	++	NT	++
DM-Cl	++	+	+	+++
MM-Cl	+	+	+	+
DM-NO$_2$	+	+	+++	++
MN-NO$_2$	+	+++	+	++
DM-COOK	+	+++	+++	+
MM-COOK	++++	++	NT	++++

*The abbreviations used to identify triazene derivatives are those used in the literature cited. NT, not tested.

Thus, although this discussion is centered on triazene derivatives that have been the subject of most studies on chemical xenogenization in vivo, the range of susceptible tumors and agents endowed with xenogenizing properties both in vivo and in vitro (14-16) allow for the contention that chemical xenogenization is not limited to selected experimental conditions but may have broader biological significance and therapeutic implications.

DRUG-MEDIATED TUMOR ANTIGENS (DMTA)

Perhaps the most impressive feature of chemically xenogenized tumors is the acquisition of novel antigens which are responsible for the new immunogenicity. The characterization of such drug-mediated antigens or DMTA has been possible through two different approaches, the first relying on reactivity of cytotoxic T lymphocytes (CTL) and the second on humoral antibody production.

1. DMTA defined by CTL responses

The presence of DMTA on DTIC-xenogenized tumors has been investigated in a number of in vitro studies in which specifically sensitized CTL were tested against ^{51}Cr-labelled target DTIC lymphomas. Thus, effector CTL can be obtained in a primary in vivo response using, as a source of lymphocytes, spleens from animals that have rejected a DTIC tumor (17). Similary, specifically cytotoxic lymphocytes are generated in vitro in a primary response (18) or in secondary responses using in vivo presensitized responder cells (19). The results of these studies are consistent with the hypothesis that DMTA are not detectable on the parental tumor and indicate that DMTA are recognized by specific CD8$^+$ lymphocytes in association with major histocompatibility complex class I determinants (20). In addition, a number of points could be firmly established that can be summarized:
    a - a DTIC tumor line, obtained through repeated in vivo exposure of parental cells to the drug, is a highly polyclonal cell population and contains a large number of variants with different antigens.
    b - Each monoclonal variant, however, is endowed with a unique set of DMTA.
    c - All xenogenized variants of a tumor retain and share tumor-associated transplantation antigens (TATA) that may pre-exist on parental cells.
    d - No cross-reactivity between any pair of xenogenized variants can be detected except for that due to parental TATA.
    e - Xenogenized variants do not display obvious changes in the qualitative expression of normal histocompatibility (H-2) antigen, as detected by cell-mediated immunity.

2. Serologically defined DMTA

More recently, we have resorted to the humoral antibody approach for the biochemical definition of DMTA. Much of the results we obtained can be thus summarized (20-22):
    a - DTIC xenogenization of a murine lymphoma will induce determinants capable of eliciting specific antibodies in immunized mice, the levels of which progressively increase if the animals are exposed to repeated sensitization. These antibodies largely belong to the IgG class.

b - The novel antigenic determinants recognized by the anti-xenogenized tumor antibody are not spatially related to the antigens shared by parental tumor, which, in turn, react predominantly with IgM present in anti-xenogenized tumor hyperimmune sera.

c - The novel determinants of the xenogenized tumor recognized by specific antibody are not expressed at detectable levels on normal cells of the same haplotype nor do they appear to be related to public specificities of selected alien histocompatibility antigens.

d - Novel antigenic determinants on xenogenized lymphoma cells can also be detected by means of monoclonal antibodies. In two different studies (22, 23), it was found that hybridomas derived from spleen cells of mice hyperimmunized against a xenogenized tumor variant may produce two kinds of antibodies: one is directed against determinant(s) shared by parental cells and nearly every xenogenized variant while the second is specific for the immunizing variant and does not cross-react with either parental cells or other xenogenized variants of that tumor.

e - Both polyclonal and monoclonal antibodies with specificity for DMTA can block induction and expression of cell-mediated immunity to xenogenized tumor cells. This suggests that the serologically defined antigens are closely related to those recognized by T cells.

f - Polyclonal anti-DMTA antibodies can be successfully exploited in the immunoprecipitation of immunogenic structures from the surface of xenogenized cells. This undoubtedly represents a major advancement towards the biochemical characterization of DMTA.

3. Immunoprecipitation of DMTA

In a recent study from our laboratory (24), a polyclonal syngeneic antiserum raised to a xenogenized tumor variant (L5178Y/DTIC) was employed in immunoprecipitation studies of cell surface and metabolically labeled tumor cells. One - and two - dimensional electrophoretic analysis of the immunoprecipitates detected a surface antigen of approximately 80 kDa. Additionally, a 45 kDa component was detected in the lysate of (^{35}S)methionine-labeled cells. A parallel immunoprecipitation was conducted using antibodies to retrovirus-related surface proteins, whose expression is known to be altered in xenogenized cells (25). Anti-xenotropic MuLV gp70 serum precipitated material whose electrophoretic pattern was similar to that of the 80 kDa surface antigen. Cross-clearing immunoprecipitation analysis revealed that the molecules reactive with the anti-DMTA antibodies were removed by the anti-xenotropic gp70 serum, whereas immunodepletion was only partial when the cell extract was first treated with the variant-specific antibodies.

PARENTAL TATA AND CROSS-PROTECTION AGAINST UNMODIFIED CELLS

In the L5178Y/DTIC tumor system, xenogenized cells can induce specific resistance to parental cells in vivo (26), generate T-dependent responses capable of conferring anti-parental tumor protection in vivo (27), and share serologically detectable TATA with parental L5178Y cells (21). Although the occurrence of DMTA on surface of xenogenized cells is believed to play a crucial role in the protection of xenogenized against parental cells, the exact mechanisms of such cross-protective immunity are poorly understood. Evidence suggest that, in addition to DMTA, the xenogenized tumor cells possess TATA in common with the parental cells, as illustrated above. The latter antigens are presumably the target of the cross-protective immune response to the parental tumor. Though it is reasonable to hypothesize an adjuvant or "helper" (28) effect of DMTA leading to a stronger anti-TATA response, recent in vitro data failed to demonstrate an increased frequency of TATA-specific cytotoxic T lymphocyte precursors (CTLp) in response to xenogenized cells. An example of this type of experiment is illustrated in Table 2, where limiting dilution (LD) microcultures of CD8$^+$ lymphocytes are assayed against parental or xenogenized cells following in vivo priming with the xenogenized tumor.

Table 2. Splenic CTLp frequencies in immunized mice*

Priming in vivo	Stimulation in vitro	CTLp frequency to:	
		L5178Y	L5178Y/DTIC
L5178Y	L5178Y/DTIC	1/41,678	1/25,623
L5178Y/DTIC	L5178Y/DTIC	1/39,842	1/5,383

*Mice received live L5178Y/DTIC or irradiated L5178Y cells 3 wk before their use as donors of CD8$^+$ responder lymphocytes in the LD microcultures.

It is interesting to note that we had previously shown that the major effector lymphocytes responsible for anti-parental tumor activity are immune Lyt-1$^+$ CD4$^+$ CD8$^-$ cells, which are capable of adoptively conferring delayed-type footpad reaction in vivo and proliferate in vitro in response to parental tumor antigens (29). In a more recent study (30), we have attempted to elucidate the cellular mechanisms underlying development of the antitumor DTH response to L5178Y cells. It was confirmed that the tumor-specific CD4$^+$ T lymphocytes are predominantly responsible for tumor-suppressive and DTH activities in our model. In agreement with previous results in other systems, these cells appeared to be sensitive to radiation in vitro, released a set of cytokines in response to parental antigens, and collaborated with radio-resistant, silica-sensitive macrophages in the host. Conceivably, the requirement for host-derived macrophages in our model serves two purposes. Accessory functions must be provided to immune CD4$^+$ lymphocytes by antigen-presenting cells. On the other hand, host-derived macrophages might act as nonspecific antitumor effector cells on activation by the lymphokines produced by tumor-immune CD4$^+$ lymphocytes. Undoubtedly, much of our results support the contention that the DTH reaction and presumably tumor inhibition in our model involve specific TATA recognition by tumor-immune CD4$^+$ cells, release of IFN gamma, and activation of effector macrophages (30).

Therefore, the presence of TATA on parental cells, although not mandatory for the induction of DMTA (31), is nevertheless necessary for the xenogenized cells to immunize effectively against parental cells. This means that the presence of DMTA makes it possible for the host to develop effective anti-TATA immunity. Although there is as yet an incomplete understanding of the effect, it seems probable that DMTA may exert an adjuvant activity through the increased function of accessory cells of the immune response and/or factors released by lymphocytes, macrophages or even the xenogenized tumor. To substantiate this hypothesis, we have recently tested a panel of murine tumors xenogenized by DTIC for production of soluble factors with lymphokine-like activity and induction of lymphokine release from naive or specifically sensitized lymphocytes (32,33). In the L5178Y tumor system, a majority of xenogenized but not parental clones were found to produce an IL-1-like factor, and this was associated, as a rule, with class II antigen expression and antigen presenting ability. However, no such properties were exhibited by the xenogenized variants of P815 and L1210Ha cells, which nevertheless occasionally expressed other lymphokine (GM-CSF, IL-3) activities. On examining the ability of xenogenized and parental tumors to cause release of IL-1, IL-2, IL-3, IFN, TNF/LT and GM-CSF from T cells, we found, as a rule, an increased lymphokine production when lymphocytes primed in vivo to a xenogenized tumor were restimulated in vitro with the same or parental (Table 3) cells.

Table 3. Lymphokine-like activity in supernatants of mixed lymphocyte/parental tumor cocultures*:

In vivo sensit.	In vitro restim.	IFN	IL-1	IL-2	IL-3	LT
L5178Y/DTIC	L5178Y	28	34	20	41	122
L1210Ha/DTIC	L1210Ha	20	0	460	70	432
P815/DTIC	P815	15	9	44	22	84

*All lymphokine activities are expressed as units/ml.

In conclusion, the data of these studies on anti-TATA reactivity substantiate the hypothesis that CD4$^+$ cells and lymphokines are crucially involved in the induction of immunity to parental tumor cells by DMTA; indeed, they speak in favor of a major involvement of these factors both in the initiation of a specific response to TATA co-expressed with DMTA (e.g., IL-1), and in the effector phase of the anti-parental tumor immunity mediated by TATA-specific CD4$^+$ cells (IFN gamma, IL-2, IL-3, LT). In contrast, increased CTL activity to parental cells does not seem to be a major mechanism of the protection induced by xenogenized against parental cells.

MAJOR HISTOCOMPATIBILITY COMPLEX (MHC) ANTIGENS ON XENOGENIZED CELLS

MHC class I and class II antigen expression on tumor cells is currently the subject of many studies of antitumor resistance, as any modification in their expression may have profound effects on the host ability to mount an effective response. In particular, it is now known that CD8$^+$ lymphocytes recognize peptide antigens in association with class I molecules, whereas CD4$^+$ cells recognize antigens presented in association with class II determinants. Therefore triazene xenogenization might affect responsiveness to both DMTA and TATA via modulation of the expression of histocompatibility antigens. In the L5178Y/DTIC tumor system, we have recently approached this issue by means of fluorescence staining with monoclonal antibodies (mAb) reactive to class I (mAb 34.1.2) or class II (mAb MK-D6) determinants of the H-2^d haplotype. In one such experiment reported in Fig. 1, it appears that

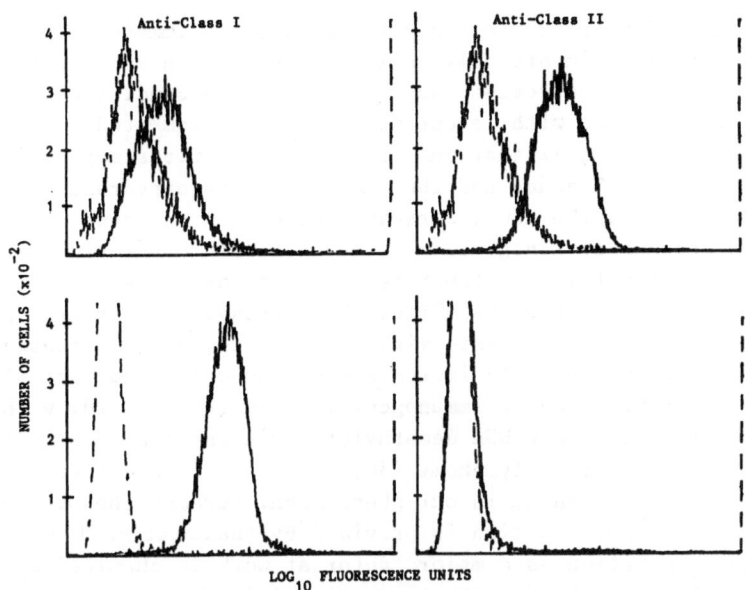

FIG. 1. Immunofluorescence profiles of L5178Y/DTIC (upper panel) and L5178Y (lower panel) reacted with monoclonal antibody to class I or class II determinants (———). (----) Background staining profiles (fluorescent antibody only).

class I antigen expression was apparently decreased in the xenogenized cells, whereas that of class II molecules was enhanced. In further studies we have investigated class I mRNA expression by Northern Blot analysis in L5178Y/ DTIC cells, and have been able to confirm the fluorescence analysis data. However, no unitary pattern of MHC antigen modification could be firmly established for the majority of the xenogenized tumors investigated. Therefore, although increased or de novo expression of class II molecules may contribute to the high immunogenicity of clonal variants of L5178Y/DTIC (32), no major role seems to played, as a rule, by changes in MHC antigen expression in the greater immunogenicity of xenogenized cells over that of the parental tumor.

MECHANISMS OF XENOGENIZATION

The finding that immunogenic tumor variants are generated after exposure of parental cells to xenogenizing chemicals raises obvious questions as to the possible mechanisms underlying the phenomenon. It should be emphasized that several points are still very controversial in this respect; however, at least in the case of triazene derivatives, the available evidence permits exclusion of some potential mechanisms. This issue has been reviewed elsewhere in detail (1). In this context, we will only comment on some recent data on the subject. A hypothesis that has always received much attention is that emergence of immunogenic tumor sublines might result from selection of pre-existing immunogenic clones. These would not be eliminated by immunologically incompetent hosts, as the DTIC-treated animals might be expected to be. Nevertheless, much available information makes this possibility rather unlikely: thus, for instance, parental lines serially transplanted in immunodepressed mice do not give rise to immunogenic variants. Furthermore, and most importantly, DTIC xenogenization can occur in the absence of drug-induced selection, and under in vitro conditions. An additional possibility, in line with recent data gathered from studies with other xenogenizing chemicals, is that the triazene derivatives may activate the expression of "silent" genes and thus condition the appearance of specificities coded for by the newly activated genes. In one such epigenetic model, the interference of xenogenizing compounds with DNA resides at the level of the enzyme that methylates the base cytosine, as it is known that the extent of cytosine methylation regulates the expression of several gene functions. Immunogenic tumor variants might, therefore, have decreased levels of methylcytosine, which would increase the transcriptional activity of genes involved in the expression of immunogenicity. In this regard, we have recently shown that no detectable DNA demethylation is associated with triazene xenogenization of a murine lymphoma (34,35). Thus, gene activation does not seem to be a major mechanism in our phenomenon. Perhaps the best explanation for drug-induced xenogenization is provided by the mutational hypothesis, which regards somatic mutation as a major factor at work in chemical xenogenization by triazenes. Despite the fact that some inconsistency exists between the actual number of histocompatibility genes and that theoretically required in the mouse to account for the observed frequency of mutational events leading to immunogenic tumor variants, this hypothesis provides the rational basis for explaining most of the experimental data available. Perhaps the mutational hypothesis could be further elaborated so as to better reconcile data on the frequency of mutation with that of variant generation. Along this line, one could hypothesize that mutagens may cause DNA rearrangements in genes forming hypermutable regions which would produce a wide variety of new antigens at very high frequencies. In this regard, it is interesting to note that retroviral DNA sequences in eukaryotic cells do represent hypermutable regions, a notion which is in agreement with our finding of DMTA as de novo expressed retrovirus-related surface proteins. However, further studies will be needed before any firm conclusion is drawn on the general mechanisms of chemical xenogenization.

CONCLUDING REMARKS

Chemical xenogenization appears to be a complex phenomenon which can be induced by agents not necessarily acting through a unitary mechanism: mutagenesis of retroviral DNA domains is perhaps the major mechanism in triazene xenogenization, but there is as yet an incomplete understanding of the phenomenon. Whatever the mechanisms of chemical xenogenization, the finding that the antigenic structure of tumor cells can be altered in vivo by employing appropriate treatments with antineoplastic agents or resorting to in vitro exposure to selected chemicals could be of relevance both for a more effective use of cytoreductive drugs and in designing new approaches to cancer immunotherapy.

On one hand, for instance, in the choice of drugs for combination chemotherapy, consideration could also be given to the xenogenizing efficiency of the drugs, or effectively xenogenizing agents might be included in the regimen.

On the other hand, entirely new immunotherapeutic approaches could be developed similar to those successfully attempted in experimental models of tumor immunotherapy (36-39). Thus, for instance:

a - Tumor-bearing hosts could be treated with xenogenizing drugs in order to increase the immunogenic potential of the malignancy.

b - The host could be treated with immunogenic tumor variants obtained in vitro through exposure of the original tumor to xenogenizing chemicals.

c - The host could be adoptively transferred with lymphocytes sensitized in vitro to immunogenic variants of the original tumor.

All of these approaches have been successfully exploited in experimental models of tumor immunotherapy using DTIC-xenogenized murine lymphomas. In one such model described by our laboratory, parental or xenogenized cells are injected into the cerebral parenchyma by means of a microsyringe. Using this model, we have been able to demonstrate that CTL raised in vivo or in vitro against a xenogenized tumor variant exerts considerable protection against the same variant when injected intracerebrally or systemically into immunodepressed hosts. More importantly, the same lymphocyte population exerts inhibition of parental tumor cells, presumably also containing TATA-specific helper/DTH lymphocytes generated in the context of the strong immunity elicited by DMTA, as illustrated above. This again suggests that otherwise nonimmunogenic TATA become the target of an effective antitumor response when coexpressed with highly immunogenic DMTA.

In conclusion, although these data are clearly a major advancement towards any practical exploitation of chemical xenogenization in the immunotherapy of human tumors, there is no doubt that much work is still needed before any firm conclusion on the possible therapeutic value of this approach can be drawn.

REFERENCES

1. Puccetti P, Romani L, Fioretti MC: Chemical xenogenization of experimental tumors. Cancer Metast Rev 6: 93-111, 1987.
2. Bonmassar E, Bonmassar A, Vadlamudi S, Goldin A: Immunological alteration of leukemic cells in vivo after treatment with an antitumor drug. Proc Nat Acad Sci (Wash) 66: 1066-1074, 1970.
3. Fioretti MC, Nardelli B, Bianchi R, Nisi C, Sava G: Antigenic changes of a murine lymphoma by in vivo treatment with triazene derivatives. Cancer Immunol Immunother 11: 283-286, 1981.
4. Riccardi C, Fioretti MC, Giampietri A, Puccetti P, Goldin A: Growth and rejection patterns of murine lymphoma cells antigenically altered following drug treatment in vivo. Transplantation 25: 63-68, 1978.
5. Campanile F, Crinò L, Bonmassar E, Houchens D, Goldin A: Radioresistant inhibition of lymphoma growth in congenitally athymic (nude) mice. Cancer Res 37: 394-398, 1977.
6. Bonmassar E, Bonmassar A, Vadlamudi S, Goldin A: Antigenic changes of L1210 leukemia in mice treated with 5-(3-3'-dimethyl-1-triazeno)imidazole-4-carboxamide. Cancer Res 32: 1446-1450, 1972.
7. Bonmassar E, Fioretti MC, Nicolin A, Spreafico F: Drug-induced modifications of tumor cell antigenicity. In Spreafico F and Arnon R (eds) Tumor associated antigens and their specific immune response. Academic Press, London, New York, San Francisco, 1979, pp 251-270.
8. Campanile F, Houchens DP, Gaston M, Goldin A, Bonmassar E: Increased immunogenicity of two lymphoma lines after drug treatment of athymic (nude) mice. J Natl Cancer Inst 55: 207-209, 1975.

9. Giampietri A, Fioretti MC, Goldin A, Bonmassar E: Drug-mediated antigenic changes in murine leukemia cells: antagonistic effects of quinacrine, an antimutagenic compound. J Natl Cancer Inst 64: 297-301, 1980.

10. Fioretti MC, Bianchi R, Romani L, Bonmassar E: Drug-induced immunogenic changes of murine leukemic cells: dissociation of onset of resistance and emergence of novel immunogenicity. J Natl Cancer Inst 71: 1247-1251, 1983.

11. Connors TA, Goddard PM, Merai K, Ross WCJ, Wilman DEV: Tumor inhibitory triazenes: structural requirements for an active metabolite. Biochem Pharmacol 25: 241-246, 1976.

12. Contessa AR, Giampietri A, Bonmassar A, Goldin A: Increased immunogenicity of L1210 leukemia following short-term exposure to 5(3,3'-dimethyl-1-triazeno)imidazole-4-carboxamide (DTIC) in vivo or in vitro. Cancer Immunol Immunother 7: 71-76, 1979.

13. Nardelli B, Contessa AR, Romani L, Sava G, Nisi C, Fioretti MC: Immunogenic changes of murine lymphoma cells following in vitro treatment with aryl-triazene-derivatives. Cancer Immunol Immunother 16: 157-161, 1984.

14. Boon T, Kellermann O: Rejection by syngeneic mice of cell variants obtained by mutagenesis of malignant teratocarcinoma cell line. Proc Natl Acad Sci USA (Wash) 74: 272-275, 1977.

15. Boon T: Antigenic tumor cell variants obtained with mutagens. Adv Cancer Res 39: 121-151, 1983.

16. Contessa AR, Bonmassar A, Giampietri A, Circolo A, Goldin A, Fioretti MC: In vitro generation of a highly immunogenic subline of L1210 leukemia following exposure to 5-(3,3'-dimethyl-1-triazeno)imidazole-4-carboxamide. Cancer Res 41: 2476-2482, 1981.

17. Nicolin A, Bini A, Coronetti E, Goldin A: Cellular immune response to a drug treated L5178Y lymphoma subline. Nature 251: 654-655, 1974.

18. Romani L, Fioretti MC, Bonmassar E: In vitro generation of primary cytotoxic lymphocytes against L5178Y leukemia antigenically altered by 5-(3,3'-dimethyl-1-triazeno)imidazole-4-carboxamide in vivo. Transplantation 28: 218-222, 1979.

19. Santoni A, Kinney Y, Goldin A: Secondary cytotoxic response in vitro against Moloney lymphoma cells antigenically altered by drug treatment in vivo. J Natl Cancer Inst 60: 109-112, 1978.

20. Romani L, Grohmann U, Fazioli F, Puccetti P, Mage MG, Fioretti MC: Cell-mediated immunity to chemically xenogenized tumors. I. Inhibition by specific antisera and H-2 association of the novel antigens. Cancer Immunol Immunother 26: 48-54, 1988.

21. Romani L, Puccetti P, Fioretti MC, Mage MG: Humoral response against murine lymphoma cells xenogenized by drug treatment in vivo. Int J Cancer 36: 225-231, 1985.

22. Grohmann U, Puccetti P, Fioretti MC, Mage MG, Romani L: Cell-mediated immunity to chemically xenogenized tumors. III. Generation of monoclonal antibodies interfering with reactivity to novel antigens. Int J Immunopharmacol 10: 803-809, 1988.

23. Testorelli C, Archetti YL, Aresca P, Del Vecchio L: Monoclonal antibodies to the L1210 murine leukemia cell line and to a drug-altered subline. Cancer Res 45: 5299-5303, 1985.

24. Grohmann U, Ullrich S, Mage MG, Appella E, Fioretti MC, Puccetti P, Romani L: Identification and immunogenic properties of an 80 kDa surface

antigen on a drug-treated tumor variant: relationship to MuLV gp70. Submitted for publication.

25. Altevogt P, Apt D: High-frequency generation of altered Mr 70 000 env glycoproteins in N-methyl-N'-nitro-N-Nitrosoguanidine. Cancer Res 48: 1137-1142, 1988.

26. Nicolin A, Spreafico F, Bonmassar E, Goldin A: Antigenic changes of L5178Y lymphoma after treatment with 5-(3,3' dimethyl-1-triazeno)imida-zole-4-carboxamide in vivo. J Natl Cancer Inst 56: 89-93, 1976.

27. Bianchi R, Romani L, Puccetti P, Fioretti MC: Inhibition of murine lymphoma growth by adoptive transfer of lymhocytes sensitized to a xenogenized tumor variant. Int J Cancer 40: 7-11, 1987.

28. Keene JA, Forman J: Helper activity is required for in vivo generation of cytotoxic T lymphocytes. J Exp Med 155: 768-782, 1982.

29. Bianchi R, Romani L, Puccetti P, Fioretti MC: Induction of tumor sup-pression and delayed-type footpad reaction by transfer of lymphocytes sensitized to a xenogenized tumor variant. Int J Cancer 42: 71-75, 1988.

30. Puccetti P, Bianchi R, Romani L, Cenci E, Fioretti MC: Delayed-type hypersensitivity to tumor antigens co-expressed with immunogenic determinants induced by xenogenization. Int J Cancer 43: 279-284, 1989.

31. Fioretti MC, Romani L, Bonmassar A, Taramelli D: Appearance of strong transplantation antigens in non-immunogenic lymphoma following drug treatment in vivo. J Immunopharmacol 2: 189-212, 1980.

32. Romani L, Grohmann U, Puccetti P, Nardelli B, Mage MG, Fioretti MC: Cell-mediated immunity to chemically xenogenized tumors. II. Evidence for accessory function and self-antigen presentation by a highly immuno-genic tumor variant. Cell Immunol 111: 365-378, 1988.

33. Romani L, Puccetti P, Grohmann U, Cenci E, Mage MG, Fioretti MC: Cell-mediated immunity to chemically xenogenized tumors. IV. Production of lymphokine activity by, and in response to highly immunogenic cells. Int J Immunpharmacol 11: 537-542, 1989.

34. Puccetti P, Fuschiotti P, Dominici P, Borri-Voltattorni C, Romani L, Fioretti MC: DNA methylating activity in murine lymphoma cells xeno-genized by triazene derivatives. Int J Cancer 39: 769-773, 1987.

35. Fuschiotti P, Fioretti MC, Romani L, Puccetti P: Lack of correlation between DNA-methylating activity and appearance of the immunogenic phenotype in clones of a murine lymphoma treated with mutagens. Cancer Immunol Immunother 29: 139-143, 1989.

36. Romani L, Bianchi R, Puccetti P, Fioretti MC: Systemic adoptive immuno-therapy of a highly immunogenic murine lymphoma growing in the brain. Int J Cancer 31: 477-482, 1983.

37. Romani L, Nardelli B, Bianchi R, Puccetti P, Mage M, Fioretti MC: Adoptive immunotherapy of intracerebral murine lymphomas: role of different lymphoid populations. Int J Cancer 35: 659-665, 1985.

38. Giampietri A, Bonmassar A, Puccetti P, Circolo A, Goldin A, Bonmassar E: Drug-mediated increase of tumor immunogenicity in vivo for a new approach to experimental cancer immunotherapy. Cancer Res 41: 681-687, 1981.

39. Romani L, Fioretti MC, Bianchi R, Nardelli B, Bonmassar E: Intracerebral immunotherapy of a murine lymphoma antigenically altered by drug treatment in vivo. J Natl Cancer Inst 68: 817-822, 1982.

THE METABOLISM OF ANTINEOPLASTIC TRIAZENES

Andreas Gescher, Lincoln L.H. Tsang and John A. Slack

CRC Experimental Chemotherapy Research Group
Pharmaceutical Sciences Institute, Aston University
Birmingham, U.K.

Is there a future for antineoplastic triazenes in the clinic? The answer is probably no as they have been under investigation now for nearly two decades without showing satisfactory promise in the treatment of human malignancies. However, 1-aryl-3,3-dialkyltriazenes are certainly very interesting experimental agents. In particular, studies of their metabolism have furnished a marked contribution to our understanding of the link between xenobiochemistry and biological activity. These studies were mostly conducted in the 60s and 70s and they established clearly that aryldialkyltriazenes require metabolism to exert their antineoplastic and carcinogenic activity (for review see ref. 1). More recently the discovery of the remarkable antineoplastic activity in mice of mitozolomide and related imidazotetrazinones, molecules in which the triazene moiety is part of a ring structure, has led to the studies of their mode of action and their metabolism[2,3]. The role which metabolism plays as a determinant of biological activity is less clear for the imidazotetrazinones than in the case of the aryldimethyltriazenes. In this overview the metabolism of antineoplastic aryldimethyltriazenes and imidazotetrazinones is compared. Such a comparison might eventually contribute to the ultimate assessment of the therapeutic benefit of treatment with these agents, if indeed there is one.

Metabolism of 1-aryl-3,3-dimethyltriazenes

The oxidative metabolism of 1-aryl-3,3-dimethyltriazenes (1) by hepatic microsomes in vitro was first reported in 1969[4]. Formaldehyde was found as a metabolite and the authors suggested that the other product of this metabolic oxidation was 1-aryl-3-methyltriazene (3). Monoalkyltriazenes such as 1-phenyl-3-methyltriazene (3) are powerful directly acting alkylating agents (Scheme 1) and carcinogens. Consequently, metabolic oxidation has been postulated to be responsible for the generation of the ultimate cytotoxicant derived from aryldialkyltriazenes, a species which causes the carcinogenicity and possibly the antitumour activity of 1-aryl-3,3-dimethyltriazenes. In most of the studies of the metabolism of aryldialkyltriazenes, which have been conducted in the wake of the original report[4], the aldehyde generated by hydroxylation of the N-alkyl groups was detected and quantitated. This route of biotransformation is mediated by hepatic mixed function oxygenase enzymes. It is only in recent years that a monomethyltriazene

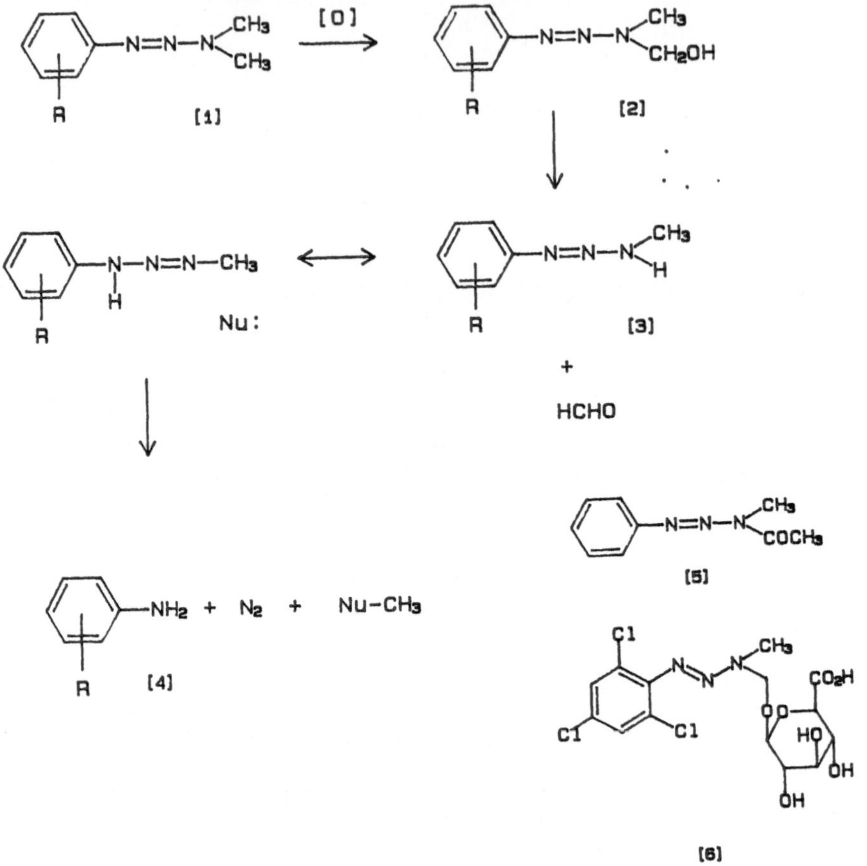

[1] → [O] → [2]

[2] → [3] + HCHO

[3] ⟷ Nu:

[5]

[4] R—NH₂ + N₂ + Nu—CH₃

[6]

has been identified unambiguously as a metabolite of a 1-aryl-3,3-dimethyltriazene: 1-(4-acetylphenyl)-3-methyltriazene (3, R=4-CH$_3$CO) together with 4-aminoacetophenone (4, R=4-CH$_3$CO) were detected as metabolites of 1-(4-acetylphenyl)-3,3-dimethyltriazene (1, R=4-CH3CO) both in the plasma of mice in vivo and in incubations with 9000Xg fractions of mouse liver homogenate in vitro[5,6]. A more indirect observation of a monomethyltriazene as a metabolite of a dimethyltriazene has been made in that one of the metabolites found in incubations of rodent hepatic microsomes with 1-phenyl-3,3-dimethyltriazene was 3-acetyl-3-methyl-1-phenyltriazene (5)[7]. Presumably the formation of this N-acyltriazene involves the monomethyltriazene (3, R=H) as a metabolic precursor which is subsequently acetylated.

Some doubt has been cast on the contention that an electrophile as indiscriminate and nonselective as a monomethyltriazene could account for the antineoplastic activity of dimethyltriazenes[8]. This doubt was based on experiments in which the cytotoxicity of a monomethyltriazene was tested in a bioassay against the TLX5 mouse lymphoma cell line which had acquired resistance to dimethyltriazenes in vivo. The monomethyltriazene was found to be nonselectively toxic to both the sensitive and resistant lines, whereas incubation of the dimethyltriazene with the 9000g fraction of a liver homogenate produced metabolites which were more toxic to the sensitive than to the resistant lymphoma cells[8]. However cytotoxic metabolites of dimethyltriazenes which are more selective than the monomethyltriazenes have never been isolated or characterised. The immediate metabolic precursor of 1-aryl-3-methyl-triazenes, the 3-(hydroxymethyl)-3-methyltriazenes (2) have been synthesised[9] and found to be equally as cytotoxic as their decomposition products, the monomethyltriazenes (3). Nevertheless such carbinolamines have not yet been identified as metabolites of aryldimethyltriazenes. 1-([3-Methyl-1-(2,4,6-trichlorophenyl)-triazen-1-yl]methyl)-beta-D-glucuronic acid (6) was identified as a metabolite in the urine of mice which had received 3,3-dimethyl-1-(2,4,6-trichlorophenyl)triazene (1,R=2,4,6 trichloro)[10]. It is likely that a carbinolamine is the precursor of the glucuronide and the authors suggested that this glucuronide may carry the monomethyltriazene to the target tissue.

The major urinary metabolite of the clinically used triazene dacarbazine (7) in man is 5-aminoimidazole-4-carboxamide (10, AIC,)[11,12]. AIC has also been identified as a product of the metabolism of dacarbazine in vitro[13,14]. From the mechanistic point of view it is difficult to envisage how AIC could be formed by a metabolic pathway other than one implicating the intermediate formation of 5-(3-methyltriazen-1-yl)imidazole-4-carboxamide (9, MTIC), followed by tautomerism and subsequent hydrolysis. The metabolic precursor of MTIC, 5-[3-(hydroxymethyl)-3-methyltriazen-1-yl]imidazole-4-carboxamide (8, HMTIC), has been characterised as a urinary metabolite of dacarbazine in rats[15]. Interestingly, this compound was more stable than MTIC in polar solvents, which suggests that HMTIC may act as a transport form of MTIC, which is the postulated antineoplastic species derived from dacarbazine.

Metabolism of imidazotetrazinones

Unlike the open-chain triazene dacarbazine, which does not degrade rapidly under physiological conditions, cyclic imidazotetrazinones undergo hydrolytic ring-opening at physiological pH. In the case of mitozolomide (11), this step is thought to yield 5-[3-(2-chloroethyl)-triazen-1-yl]imidazole-4-carboxamide (12) as an intermediate[16], which may well be the ultimate antineoplastic metabonate of the parent drug[2]. Hydrolysis of temozolomide (13), the N-methyl analogue of mitozolomide, should furnish MTIC (9), which is the cytotoxic and short-lived major metabolite of dacarbazine, as outlined above. When temozolomide was incubated in RPMI 1640 medium including horse serum, MTIC was found by cochromatography with authentic material on analysis of the incubation mixture by HPLC (Tsang, Gescher and Slack, unpublished). Likewise, MTIC has also been characterised in the plasma of mice 5min after they had

CONH₂ ... N=N ... CH₂CH₂Cl

$CONH_2$ N—N N CH₂CH₂Cl O [11]

$CONH_2$ N=N NH H—N CH₂CH₂Cl [12]

$COOH$ N—N N CH₃ O [14]

$CONH_2$ N=N N CH₃ O [13]

O=C—N(CH₃)₂ N—N N CH₂CH₂Cl O [15]

O=C—N(CH₃)(H) N—N N CH₂CH₂Cl O [16]

received temozolomide. There is therefore no doubt about the contention that temozolomide is a prodrug of MTIC. Treatment with temozolomide may offer the advantage over dacarbazine in that the formation of the active species MTIC is more predictable in the case of temozolomide as it does not require metabolism.

Studies on the cytotoxicity of antineoplastic imidazotetrazinones have clearly shown that metabolism is not a prerequisite for cytotoxicity. In view of the chemical properties of mitozolomide and temozolomide it seems important to address the question: are these molecules metabolised or is the rate of their chemical decomposition much faster than the rate of generation of any metabolites? In a recent study two metabolites were found in the plasma of patients who were treated with temozolomide during a clinical trial at the Queen Elizabeth Hospital in Birmingham (Tsang, Quarterman, Farmer, Slack and Gescher, unpublished). One of the metabolites was identified by mass spectrometry and high-field ^{1}H-NMR spectroscopy as the imidazocarboxylic acid derivative (14) of temozolomide (13) generated by hydrolysis of the carboxamide function. The chemical nature of the other metabolite remains unsolved. It possesses the UV spectroscopic characteristics of triazenes, contains the unchanged imidazocarboxamide structure according to its ^{1}H-NMR spectrum, and it is very water soluble. The ^{1}H-NMR spectrum of the isolated metabolite suggests also that the molecule has lost the N-methyl moiety contained in the parent temozolomide. This metabolite, which was also found in the plasma of mice which had received temozolomide, has acidic properties and seems to be a detoxification product of temozolomide, as it lacks pronounced cytotoxicity when incubated with murine TLX5 lymphoma cells.

Following the synthesis of mitozolomide almost a decade ago a large number of analogues were synthesised in order to help with the understanding of the chemical features of the drug molecule which are essential for its cytotoxic and antineoplastic properties. Substituents in position 8 have been shown to be important determinants of cytotoxicity. Analogues bearing a carboxamide group on carbon 8 of the imidazotetrazinone ring require a N-H moiety for maximum cytotoxity. The dimethylcarbamyol analogue (15) of mitozolomide was markedly less cytotoxic than mitozolomide or its monomethyl analogue (16)[3]. Nevertheless, the antitumour activity of the three agents in mice bearing the TLX5 lymphoma was similar. This observation can be explained by the fact that dimethylmitozolomide (15) undergoes metabolism in vivo to the potently cytotoxic monomethylmitozolomide (16). In accordance with this hypothesis incubation of TLX5 cells with dimethylmitozolomide in the presence of murine hepatic microsomes brought about a significant increase in cytotoxicity, probably because dimethylmitozolomide underwent metabolic activation by N-demethylation generating methylmitozolomide. Methylmitozolomide (16) was indeed characterised as a metabolite of dimethylmitozolomide (15) under these incubation conditions[3]. This finding is consistent with the suggestion that methylmitozolomide may contribute to the cytotoxicity of dimethylmitozolomide in vivo.

Conclusion

        The evidence for the involvement of metabolic oxidation in the mechanism of antitumour activity of aryldimethyltriazenes is overwhelming. However several crucial mechanistic details which link the metabolism of aryldimethyltriazenes with their antineoplastic activity remain obscure. Answers to the following three questions appear pivotal in the elucidation of this link:

(i)   How can the aggressively reactive monomethyltriazene generated by metabolism of a dimethyltriazene cause selective cytotoxicity?

(ii)   Is the metabolic route which leads to the antineoplastic species identical with that which generates the carcinogenic metabolite?

(iii) Are differences in the metabolism between dimethyltriazenes and dialkyltriazenes with alkyl groups other than methyl responsible for the marked difference in antineoplastic activity observed in some murine tumours?

        The role of metabolism in the causation of biological activity does not seem to be as crucial for the antineoplastic imidazotetrazinones than in the case of the aryldimethyltriazenes. Nevertheless fine-tuning of cytotoxic activity of the imidazotetrazinones seems to be possible by exploitation of the knowledge of structural features of these molecules which are prone to be metabolised.

Acknowledgment

        We thank Dr. C. Quarterman (Aston University) and Dr. P.B. Farmer (MRC Toxicology Unit, Carshalton) for their contribution to the work on temozolomide. The work was supported by the Cancer Research Campaign of Great Britain.

References

1. A. Gescher and M.D. Threadgill, The metabolism of triazene antitumor drugs, Pharmac. Ther. 32: 191 (1987).

2. C.M.T. Horgan and M.J.Tisdale, Antitumour imidazotetrazines. IV. An investigation into the mechanism of antitmumour activity of a novel and potent antitumour agent, mitozolomide. Biochem. Pharmacol. 33: 2185 (1984).

3. K.R. Horspool, C.P. Quarterman, J.A. Slack, A. Gescher, M.F.G. Stevens and E. Lunt, Metabolic activation and murine pharmacokinetics of the 8-(N,N-dimethylcarboxamide) analog of the experimental antitumor drug mitozolomide (NSC353451), Cancer Res., 49: 5023 (1989).

4. R. Preussman and A. von Hodenberg, Mechanism of carcinogensis of 1-aryl-3,3-dialkyltriazenes. Enzymatic dealkylation by rat liver microsomal fraction in vitro, Biochem. Pharmacol. 18: 1 (1969).

5. P. Farina, A. Gescher, J.A. Hickman, J.K. Horton, M. D'Incalci, D. Ross, M.F.G. Stevens and L.C. Torti, Studies of the mode of action of antitumour triazenes and triazines. IV. The metabolism of 1-(4-acetylphenyl)-3,3- dimethgyltriazene. Biochem. Pharmacol. 31: 1887 (1982).

6. P. Farina, E. Benfenati, R. Reginato, L. Torti, M. D'Incalci, M.D. Threadgill and A. Gescher, Metabolism of the anticancer agent 1-(4-acetylphenyl)-3,3- dimethyltriazene, Biomed. Mass Spectrom., 10: 485 (1983).

7. B.L. Pool, Microsomal mediated metabolism of dialkylaryltriazenes. II. Isolation and identification of metabolites of 3,3-dimethyl-1-phenyltriazene, J. Cancer Res. Clin. Oncol. 93: 221 (1979).

8. A. Gescher, J.A. Hickman, R.J. Simmonds, M.F.G. Stevens and K. Vaughan, Studies of the mode of action of antitumour triazenes and triazines. II. Investigation of the selective toxicity of 1-aryl-3,3,-dimethyltriazenes. Biochem. Pharmacol. 30: 89 (1981).

9. A. Gescher, J.A. Hickman, R.J. Simmonds, M.F.G. Stevens and K. Vaughan, Alpha-hydroxylated derivatives of antitumour dimethyltriazenes, Tetrahedron Lett. 50: 5041 (1978).

10. G.F. Kolar and R. Carubelli, Urinary metabolites of 1-(2,4,6-trichlorophenyl)-3,3,-dimethyltriazene with an intact diazoamino structure, Cancer Lett. 7: 209 (1979).

11. G.E. Householder and T.L. Loo, Elevated urinary excretion of 4-amino imidazole-5-carboxamide in patients after iv injection of 4-(3,3-dimethyl-1-triazeno)imidazole- 5-carboxamide, Life Sci. 8: 533 (1969).

12. J.L. Skibba, G. Ramirez, D.D. Beal and G.T. Bryan, Metabolism of 4(5)-(3,3-dimethyl-1-triazeno)-imidazole-5(4)-carboxamide to 4(5)-amino-imidazole-5(4)carboxamide in man, Biochem. Pharmacol. 19: 2043 (1970).

13. N.S. Mizuno and E.W. Humphrey, Metabolism of 5-(3,3-dimethyl-1-triazeno)imidazole-4-carboxamide (NSC 45388) in human and tumor tissue, Cancer Chemother. Rep. 56: 465 (1972).

14. D.L. Hill, Microsomal metabolism of triazenylimidazoles, Cancer Res. 35: 3106 (1975).

15. G.F. Kolar, M. Maurer and M. Wildschütte, 5-(3-Hydroxymethyl-3-methyltriazeno)imidazo-4-carboxamide, a metabolite of dacarbazine (DIC, DTIC, NSC 45388), Cancer Lett. 10: 241 (1980).

16. M.F.G. Stevens, J.A. Hickman, R. Stone, N.W. Gibson, E. Lunt, C.G. Newton and G.U. Baig, Antitumor imidazotetrazines. I. Synthesis and chemistry of 8-carbamoyl-3-(2-chloroethyl)imidazo[5,,1-d]-1,2,3,5-tetrazine-4(3H)-one, a novel broad spectrum antitumor agent, J. Med. Chem. 27: 196 (1984).

NOTES ON THE METABOLISM, PHARMACOKINETICS   AND MODE OF ACTION OF
N-METHYL AND N-ETHYL-TRIAZENES IN RELATION TO THEIR PHARMACOLOGICAL
ACTIVITY

Maurizio D'Incalci, Tina Colombo, Pierluigi Farina
Cecilia Mannironi, Pietro Taverna and Carlo V. Catapano

Istituto di Ricerche Farmacologiche "Mario Negri"
Via Eritrea 62, 20157 Milan, Italy

INTRODUCTION

Dacarbazine (DTIC) is an antitumor agent which has a well documented activity against some human tumors [1,2]. Its preclinical and clinical activity is not striking, but is justified by its relatively low toxicity. As for most anticancer agents the molecular mode of action of dimethyltriazenes is still debatable but most authors recognize the importance of N-demethylation for the activation of these drugs [3,4]. The monoalkylderivative is probably the alkylating species responsible for the antitumor activity. Structure activity studies have essentially failed to find any new dimethyltriazene clearly superior to those already available. However, several studies   have shown that diethyltriazenes are inactive. The reasons for the discrepant activity of dimethyltriazenes and diethyltriazenes is still unknown.

In this paper we discuss a series of studies on the pharmacological properties of dimethyltriazenes and diethyltriazenes in order to get some clue to the profound differences in the activity of these compounds and to stress specific points   and open questions for critical evaluation.

The optimistic thinking behind many studies shown here was that a better understanding of the reasons for the higher antitumor selectivity of dimethyltriazenes could provide a handle to start to elucidate the biochemical and pharmacological basis of selectivity.

MATERIALS AND METHODS

Animals and tumors. L1210 cells were transplanted i.p. in CD2F1 male mice (10^5 cells/mouse); T cell lymphoma EL-4 were transplanted i.p. in C57 Bl   female mice. The antitumùor activity was evaluated by recording the survival time as described by Geran et al.[5].

Drugs. The various triazenes were supplied either by Prof MFG Stevens and Dr Gescher of Aston University   or by Prof Nisi and Dr Lassiani of University of Trieste.

Metabolism and pharmacokinetics

Mice livers were homogenized in ice-cold 0.15 M KCl buffer solution (pH 7.4) in a ratio of 1:5 (W:V). The homogenates were centrifuged   at 9,000 x g for 20 min. Incubations were started by adding the triazene

Triazenes, Edited by T. Giraldi *et al.*
Plenum Press, New York, 1990

in 0.1 ml acetone to the supernatant fraction (1 ml) with NADPH (2 mg) as cofactor dissolved in 0.8 ml 0.15 M KCl buffer solution. Incubations were made in the presence of atmospheric oxygen. At different intervals 1 ml aliquots of the incubation mixtures were deproteinised by addition of 1ml ice-cold acetone. Samples were centrifuged at 600 g for 3 min then injected into a HPLC column. Further details are described in previous papers of this laboratory[6,7,8].

For in vivo determinations 4 mice for each time intervals were sacrificed, and drug determinations were performed in plasma with ananlytical methods previously described in details[9]. The area under the curve of drug concentration vs time (AUC) values were calculated by the trapezoidal method.

Cytotoxicity evaluation

L1210 and L1210/BCNU cells were seeded at 0.2×10^6 cells/ml and incubated for 24h. The cultures were drug treated for 1h at 37°C, washed in PBS by centrifugation and resuspended in fresh medium or in medium supplemented with 2mM 3-aminobenzamide (3AB). Cells were counted at the required times using a Coulter Counter.

Evaluation of DNA-breaks

In order to evaluate DNA-damage we used the method of alkaline elution according to the method of Kohn recently described in detail[10].

RESULTS AND DISCUSSION

Antitumor activity

As can be seen in table 1 1-(4-acetylphenyl) - 3,3-dimethyltriazene (p-Ac(CH3)$_2$) has a moderate but significant activity against L1210 and EL-4 mouse tumors. In contrast its diethyl analogue p-Ac(C2H5)$_2$ was very toxic and without any detectable antitumor activity. Daily i.p. doses of 5mg/Kg p-Ac(C2H5)$_2$ or 20mg/Kg p-AC(CH3)$_2$ on days 1 to 6 were approximately equitoxic as assessed by determining animals' weight, but only the dimethyltriazene showed antitumor activity. We cannot exclude that also p-Ac(C2H5)$_2$ has some potential antitumor activity at higher doses but certainly the therapeutic index is much lower than that of p-AC(CH3)$_2$. The much greater therapeutic index of p-Ac(CH3)$_2$ than p-Ac(C2H5)$_2$ was observed on M5076 reticulum cell sarcoma of the mouse too. On this experimental model the dimethyltriazene showed antitumor and antimetastatic effects whereas the dimethyltriazene was therapeutically inactive but very toxic[7].

The higher toxicity of p-Ac(C2H5)$_2$ than of p-Ac(CH3)$_2$, evaluated as weight loss and white blood cell count drop was also seen in normal (i.e non tumor bearing) C57 Bl mice (data not shown) and it is therefore not related to the presence of the tumor.

In table 1 some data on the activity of N-dealkyl derivatives are reported. P-Ac CH3 showed a moderate but detectable antitumor activity, whereas p-Ac C2H5 was inactive. In interpreting these data, however, it is important to consider that the stability of p-Ac C2H5 is much lower than that of p-Ac CH3 (see next paragraph) and it is thus possible that p-Ac C2H5 is decomposed before reaching the critical molecular target at effective concentrations.

Other experiments corroborate the view that the antitumor activity of dimethyltriazenes is mediated by the monomethylderivative. For

TABLE 1. Antitumor activity of 1-(4-acetylphenyl)-3,3-dimethyltriazene(p-Ac (CH3)2),1-(4-acetylphenyl)-3,3-diethyltriazene (p-Ac (C2H5)2) and their N-dealkylated metabolites on L1210 and EL-4 mouse tumors.

DRUG	TREATMENT SCHEDULE (mg/Kg x days)[a]	TUMOR L1210 (T/C%)[b]	EL-4 (T/C%)
p-Ac (CH$_3$)$_2$	20x6	125	150
p-Ac CH$_3$	20x6	131	–
p-Ac (C$_2$H$_5$)$_2$	20x4[c]	–	58
	5x6	113	117
p-Ac C$_2$H$_5$	20x6	106	–

[a]All treatments were started on day 1 after tumor transplant and were given by the i.p. route. Each group consisted of 10 mice.
[b]T/C % = median survival time of treated mice / median survival time of untreated controls x 100.
[c]Treatment was stopped after 4 days because animals were clearly suffering from drug treatment.

example, the data shown in table 2 indicate that the methyltriazene formed from N-demethylation of 1-p-(3,3-dimethyl-1-triazeno) benzoic acid potassium salt is as active as the parent drug.

The finding that dimethyltriazenes are active antitumor agents whereas diethyltriazenes are inactive but still toxic (e.g. causing a dramatic drop in white blood cells) stimulated the interest of our and other laboratories[7]. It was in fact possible that this research could provide a hint in the partial selectivity of dimethyltriazenes.

TABLE 2. Antitumor activity of 1-p-(3,3-dimethyl -1-triazeno) benzoic acid potassium salt (DM-COOK) and its metabolite (MM-COOK) on L1210 leukemia.

DRUG	TREATMENT SCHEDULE (mg/Kg x days)[a]	L1210 (T/C %)[b]
DM-COOK	100 x 6	175
	200 x 1	150
	400 x 1	169
MM-COOK	20 x 6	175

[a]Treatments were started on day 1 after tumor i.p. transplant and were also given i.p. .
[b]T/C % = median survival time of treated mice /median survival time of untreated controls x 100.
Each group consisted of 10 mice.

p-Ac(CH3)$_2$ and p-Ac(C2H5)$_2$ were selected as model drugs for these studies[7]. The analytical procedures that we developed allowed us to determine the parent drugs, the N-dealkylderivatives (i.e. monomethyl or monoethyl) and 4-amino acetophenone resulting from the further metabolism or decomposition of the N-dealkylderivatives .The HPLC assays were sufficently sensitive and specific to be employed for both *in vitro* and *in vivo* experiments.

We investigated the *in vitro* stability and metabolism of p-Ac(CH3)$_2$ and of p-Ac(C2H5)$_2$ of their monoalkylderivatives and of 4-aminoacetophenone. We could not determine the methylol or ethylol formed as intermediates of the oxidative N-dealkylation, because these metabolites were unstable and were rapidly converted to their respective N-monoalkyltriazenes. The rate of decomposition and metabolism of the two monoalkylderivatives was calculated experimentally . For the metabolism we used mouse liver extracts (supernatants of 9,000 x g) and cofactors. At physiological pH p-Ac CH3 was chemically more stable than p-Ac C2H5 and both compounds were found to undergo both spontaneous degradation and metabolism. Monoalkyltriazenes were converted almost completely to 4-aminoacetophenone. A mathematical model has been used to elaborate the data taking into account that while monoalkyltriazene was formed it was concomitantly degraded and metabolised. The relatively high stability of dialkyltriazenes and of 4-aminoacetophenone allowed us to determine at any time the amount of dialkyltriazene which was metabolised to its monoalkylderivative. By applying this linear mathematical model it was possible to calculate that approximately 80% of the initial concentration of p-Ac(CH3)$_2$ was biotransformed to p-Ac CH3, whereas less than 30% of the initial substrate concentration of p-Ac (C2H5)$_2$ was metabolised to p-Ac C2H5.

This difference can be particularly relevant considering that the monoalkylderivatives are the alkylating species presumably responsible for the antitumor activity of alkyltriazenes. The metabolic differences of p-Ac (CH3)$_2$ and p-Ac (C2H5)$_2$ was confirmed also *in vivo*, in mice treated with the two drugs i.p. at doses of 40 mg/Kg. The AUC of p-Ac (CH3)$_2$ and of its metabolites p-Ac CH3 and 4-aminoacetophenone were 440,551 and 433 µg/mlxmin respectively, whereas the AUC of p-Ac (C2H5) and of its metabolites p-Ac C2H5 and 4-aminoacetophenone were 170, 49 and 275 µg/mlxmin respectively. Therefore the low antitumor activity of p-Ac(C2H5) could be related to the relatively low plasma levels of p-Ac C2H5, which is probably the active metabolite. On the other hand the high toxicity of p-Ac (C2H5)$_2$ as compared to p-Ac (CH3)$_2$ does appear related to the levels of the N-dealkylated metabolite. It is suggested that p-Ac (C2H5)$_2$ undergoes some other metabolic pathways leading to some very toxic metabolite (e.g. N-oxide ?) which does not possess antitumor activity. Alternatively the low levels of p-Ac C2H5 could be due to the rapid covalent binding of this metabolite to plasma or tissue proteins which makes it inextractable with organic solvents. The available experimental data are not sufficent to indicate which of the two is the most likely hypothesis.

Although the selection of two model drugs for our studies appeared a sensible approach we cannot by any means generalise what we have found to other dimethyl or diethyltriazenes. From further studies with other aryldimethyltriazenes we have in fact realised that slight changes in the chemical structures of these drugs can change dramatically their pharmacological behaviour. An example is illustrated by analysing the metabolism and pharmacokinetic properties of p-(3,3-dimethyl-1-triazeno)-benzoic acid potassium salt (pCOOK-DMT). This compound is still at the first stages of clinical investigation and will be further discussed by Newell et al. (see chapter...). We have investigated the metabolism of this compound both *in vitro* and *in vivo*. *In vitro*, in presence of liver extracts and cofactors p-COOK-DMT appeared to undergo N-demethylation, but at a very low rate. Small amount of formaldehyde were found by Sava et al.[11,12] by incubating p-COOK-DMT with liver extracts. After 30 min of incubation with the supernatant of 9,000 x g fraction of mouse liver only 10 % of the initial substrate concentration of pCOOK-DMT was metabolised, whereas in the same experimental condition more than 90% of p-Ac (CH3)$_2$ was biotransformed.

In the incubation mixture the extracted N-monomethylderivative (pCOOK-MMT) of pCOOK-DMT was so little that it was hardly identifiable[9]. Also in plasma of mice receiving therapeutic doses of pCOOK-DMT the levels of pCOOK-MMT were undetectable with an assay which achieved the sensitivity of 100 ng/ml of plasma. An interesting property shown by pCOOK-MMT was its very high stability in plasma [9]. It was found that after 80 min of incubation with mouse plasma at 37° the concentration of pCOOK-MMT was unchanged. After the same incubation time approximately 50% of pCOOK-MMT was decomposed when incubated in a solution of 4% albumin, whereas 100% was decomposed after 20 min in 0.15M potassium chloride buffer. These data suggest that plasma proteins (albumin and others) make pCOOK-MMT highly stable. This unexpected stability of pCOOK-MMT in plasma is probably of great importance to understand the pharmacological activity of pCOOK-DMT. It is in fact conceivable that the low amount of pCOOK-MMT formed is not decomposed in plasma and can reach the tumor target cells. It is therefore likely that although the peak level of pCOOH-MMT is very low because of the high stability of his high stability its AUC is sufficiently high to exert antitumor effects. We do not want to discuss the distribution and metabolism of pCOOK-DMTin mice too extensively because we have recently described it in detail [8], but we would like to stress that what has been suggested by examining the results of the experiments performed *in vitro* and *in vivo* in mice is consistent with what has been found by Newell et al during the early clinical investigations (see the chapter by Newell et al.) . Their data in fact suggest that the pharmacological activity of pCOOK-DMT is related to the plasma levels (plasma AUC values) of pCOOH-MMT.

The example given on the peculiar pharmacological properties of pCOOK-DMT suggests caution in extrapolating the results obtained with a single aryltriazene to others even if they are structurally very similar.

Mechanism of action

The complexity of the metabolism of dimethyl and diethyltriazenes make the studies on the molecular mechanism of action of these drugs particularly difficult. It is infact necessary to carry out the experiments in presence of a metabolic system with the simultaneous evaluation of the metabolic products, their further metabolism and decomposition and their reaction with the molecular targets.

Three-substituted imidazotetrazinones [13] have not been of interest only because of their possible clinical utility (see the chapter by Newland et al.), but have also been important as tools to investigate the mechanism of action of methyl and ethyltriazenes. Eight-carbomoyl - 3-(2-methyl) imidazo (5,1d)-1,2;3,5 - tetrazin - 4-(3H)-one (Temozolomide) or the 3-ethyl analogue (ethazolastone), synthetised at Aston University, have been selected by Tisdale's laboratory and by our own laboratory for some mechanistic studies. Although these two compounds can also be metabolised in very small amounts (see the chapter by Gescher et al.) it appears that both componds *in vitro* (at phisiological pH), or *in vivo* decompose spontaneously to 5-(3-methyl-1-triazeno) imidazole- 4-carboxamide (MTIC), which is supposed to be the active metabolite of the antitumor agent dacarbazine

(DTIC), or to the ethyl analogue respectively. As expected temozolomide is an active antitumor agent. In addition temozolomide showed also the ability to differentiate human K 562 erytroleukemia cells[14,15]. In contrast ethazolastone was inactive either as antitumor cytotoxic agent or as a differentiating agent[14].

Before undergoing comparative experiments on the ability of temozolomide and ethazolastone to cause DNA damage we evaluated the rate of decomposition of both compounds in medium RPMI supplemented with 10% FCS (i.e. the culture medium in which the molecular pharmacology experiments were carried out). The rate of decomposition reproduces the rate of formation of the alkylating species and is linear at different concentrations. Fig 1 shows a representative experiment which indicates that ethazolastone is decomposed about twice as slow as temozolomide. Therefore the experiments on DNA damage and cytotoxicity were done using a concentration of ethazolastone twice that of temozolomide. L1210 mouse leukemia cell line and a subline (L1210/BCNU), which was found partially resistant to methyltriazenes and to have higher levels of O^6 alkylguanine-DNA alkyltransferase (ATase) than the parent line [16], were used. The aim of these studies was to evaluate the amount of DNA damage (DNA breaks and alkali labile sites) caused by temozolomide and ethazolastone and the rate of repair of the DNA lesions.

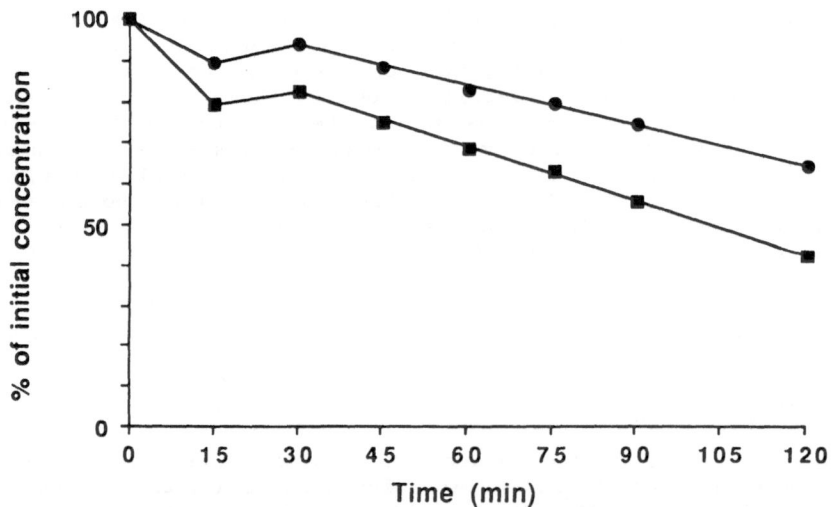

Fig.1. Decomposition of temozolomide (■) or ethazolastone (●) in RPMI 1640 culture medium supplemented with 10% FCS at 37°C. The initial concentrations were 100 μg/ml for each drug. Temozolomide and ethazolastone were assayed with the HPLC method described by Slack et al.[22]

Fig 2 shows the alkaline elution profile of DNA of L1210 and L1210/BCNU cells exposed to temozolomide or ethazolastone. It appears that in both cell lines temozolomide caused much more DNA breaks (i.e. a faster elution rate of DNA) than ethazolastone. The DNA breaks observed can be probably more appropriately defined as DNA alkali sites[16] because of the shape of the curves (i.e.curved downward). The higher number of DNA-breaks produced by temozolomide could be due to a higher number of alkylations to N7 guanine, which is the main site of alkylation. However, this hypothesis contrasts with the results of the studies reported by Bull and Tisdale who found a similar number of alkylations in DNA of two human cancer cell lines exposed with

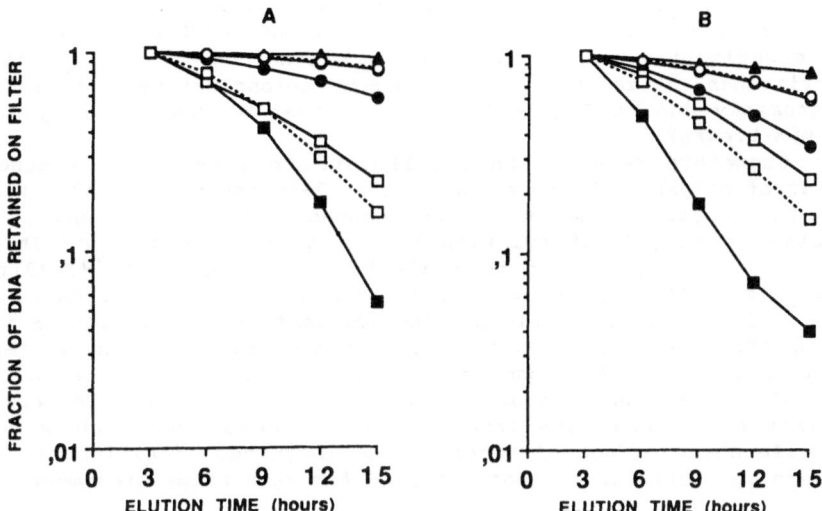

Fig.2 DNA-single strand breaks level in L1210 (panel A) or L1210/BCNU (panel B) after 1h treatment with 400µM temozolomide (■□) or 800µM ethazolastone (● O). Alkaline elution experiments were performed immediately after drug treatment (closed symbols) or after 24 hours recovery (open symbols) in culture medium with (----) or without (____) 2 mM 3-aminobenzamide. (▲) control cells.

radioactive temozolomide or ethazolastone[17]. An alternative hypothesis is that the higher number of DNA-breaks after temozolomide than after ethazolastone could be related to a greater activity of repair enzymes. For example glycosilases might remove more methylated than ethylated bases, thus generating more base free sites (which at alkaline pH are converted into DNA-breaks) after temozolomide than after ethazolastone treatment. The experimental data obtained thus far do not proove or disprove this possibility. However, although a considerable amount of DNA damage produced by either drug in both L1210 and L1210/BCNU was repaired in the 24 h following drug exposure, DNA-breaks were still higher in cells exposed to temozolomide than those exposed to ethazolastone and this would not indicate a more efficient repair of temozolomide-induced DNA damage. Probably the quality of DNA lesions is different in cells treated with temozolomide or ethazolastone and this accounts for the different pharmacological activity of the two compounds.

We did not determine the alkylations to O^6 guanine induced by temozolomide or ethazolastone, though we know that a reasonable explanation for the differential sensitivity of L1210 and L1210/BCNU to methyltriazenes is the different ATase content in these cell lines (see chapter by Margison et al.). However also supposing that ethazolastone produces less alkylations of 0^6 guanine this would not explain the greater number of DNA-breaks caused by temozolomide than by ethazolastone since the alkylation of 0^6 guanine (and also its repair by ATase) do not generate DNA-breaks. In addition the repair of 06methylguanine is reportedly [18,19] more efficient than that of 0^6 ethylguanine, thus if a differential alkylation of 0^6 guanine by temozolomide and ethazolastone were important for the cytotoxicity one would predict that ethazolastone would be more active. That is clearly not the case. Therefore the differential cytotoxicity and antitumor activity of temozolomide and ethazolastone is probably unrelated to the alkylations of 0^6 guanine. For this reason L1210/BCNU seemed to us a useful experimental model. In this cell line some cytotoxicity of methyltriazenes at relatively high concentrations was evident and this was presumably due to DNA lesions other than 0^6 guanine alkylations (e.g. DNA-breaks).

We therefore selected this cell line to investigate whether an inhibitor of polyADP ribolsylation such as 3-aminobenzamide (3-AB) could affect the repair of the DNA lesions caused by the two compounds. The inhibition of polyADPribosylation has been reported to inhibit the repair of DNA damage caused by radiation or by chemicals[20]. Although the exact role of polyADPribosylation in DNA repair mechanisms is still unclear[21], it could indirectly provide information on differences in the nature of DNA damage produced by different chemicals and their repair. As shown in fig 2, 3-AB caused some delay in the repair of temozolomide induced DNA breaks, whereas did not produce any effect on the repair of ethazolastone-induced DNA-breaks. Fig 3 shows that the moderate antiproliferative effect of temozolomide increased significantly when cells were incubated in presence of 3-AB for 24 h after treatment.

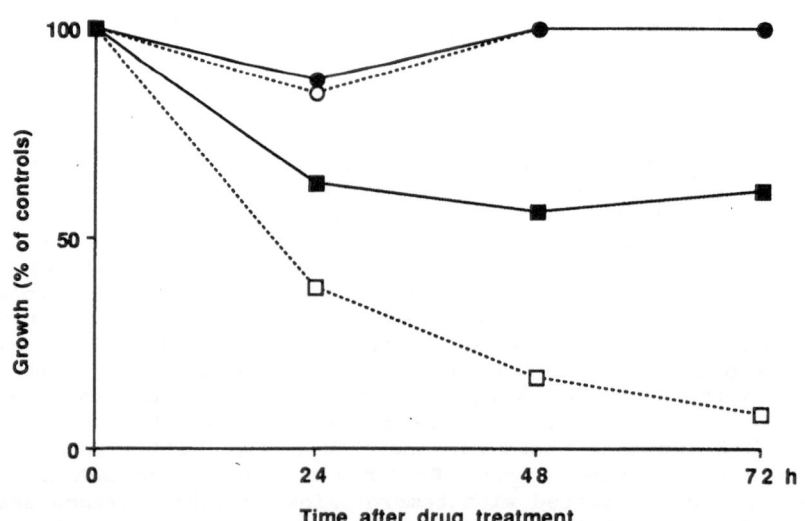

Fig.3.Growth inhibition of L1210/BCNU exposed to 400 μM temozolomide (■ □) or to 800 μM ethazolastone (● O) for 1h. (⎯⎯) after drug treatment, cells were washed and mantained in drug free medium. (----) after drug treatment, cells were washed and mantained in medium containing 2mM 3-aminobenzamide. Each value was the mean of 4 determinations and S.D. was always less than 10%.

On the other hand ethazolastone alone or also with 3-AB was inactive. These data suggest differences in the nature of DNA damage produced by temozolomide and ethazolastone. Other studies are necessary to elucidate these differences in detail.

CONCLUSIONS

The studies shown in the present paper have underlined several differences between dimethyl and diethyltriazenes. Dimethyltriazenes are effective anticancer agents with relatively low toxicity, whereas diethyltriazenes are inactive as antitumor agents but are very toxic. Using a dimethyltriazene and a diethyltriazene as model compounds we have found marked differences in their metabolic and pharmacokinetic features. However, different aryltriazenes possess different metabolic properties and it is therefore difficult to generalize the data obtainedfor one triazene to all triazenes. In addition other experiments conducted with substituted imidazotetrazinones which spontaneously decompose to methyl or to ethyl triazenes indicate differences in DNA damage caused by these compounds and of their repair.

Further research is certainly needed to elucidate the qualitative differences of the molecular interactions of methyltriazenes and ethyltriazenes with cellular macromolecules, which are presumably the basis of the profound differences in their pharmacological properties.

REFERENCES

1. J.K. Luce, W.G. Thurman, B.L. Isaacs, and R.W. Talley, Clinical Trials with the antitumor agent 5-(3,3-dimethyl-1-triazeno) imidazole-4-carboxamide (NSC 45388), Cancer Chemother. Rep. 54: 119 (1970).

2. R.L. Comis, DTIC (NSC 45388) in malignant melanoma: a perspective, Cancer Treat. Rep. 60: 165 (1976).

3. A.H. Gerulath, and T.L. Loo, Mechanism of action of 5-(3,3-dimethyl-1-triazeno)imidazole-4-carboxamide in mammalian cells in culture, Biochem.Pharmacol. 21: 2335 (1972).

4. N.S. Mizuno, R.W. Becker, and B. Zakis, Effects of 5-(3-methyl-1-1triazeno)imidazole-4-carboxamide (NSC 407347), An alkylating agent derived from 5-(3,3-dimethyl-1-triazeno) imidazole-4-carboxamide (NSC 45388), Biochem. Pharmacol. 24: 115 (1975).

5. R.I. Geran, N.H. Greenberg, M.M.Macdonald, A.M. Schumacher, and B.J. Abbott. Protocols for screening chemical agents and natural products against animal tumors and other biological systems, Cancer Chemother.Rep. 3: 1 (1972).

6. P. Farina, A. Gescher, J.A. Hickman, J.K. Horton, M. D'Incalci, D. Ross, M.F.G. Stevens, and L. Torti, Studies of the mode of action of antitumour triazenes and triazines- IV. The metabolism of 1-(4-acetylphenyl)-3,3- dimethyltriazene, Biochem.Pharmacol. 31: 1887 (1982).

7. P. Farina, L. Torti, R. Urso, J.K. Horton, A. Gescher, and M. D'Incalci, Comparison of metabolism and activity of an aryldimethyltriazene and an aryldiethyltriazene, Biochem. Pharmacol. 35: 209 (1986).

8. E. Benfenati, P. Farina, T.Colombo, G. De Bellis, M.V.Capodiferro, and M. D'Incalci, Metabolism and pharmacokinetics of *p* - (3,3-dimethyl-1-triazeno)benzoic acid in M5076 sarcoma - bearing mice, Cancer Chemother. Pharmacol. 24: 354 (1989).

9. P. Farina, E. Benfenati,L. Lassiani, C. Nisi, and M. D'Incalci, High-performance liquid chromatographic assay for the determinatio of *p* -(3,3-dimethyl-1-triazeno)benzoic acid in mouse plasma, J. Chrom. Biomed. Applications 345: 323 (1985).

10. K.W. Kohn, R.A.G. Ewig, L.C. Erickson, and L.A. Zwelling, Measurement of strand breaks and cross-links by alkaline elution, In: "DNA repair.A laboratory manual of research procedures. Vol 1, part B" E.C. Friedberg and P.C. Hanawalt, ed., Marcel Decker Inc., New York and Basel, (1981).

11. G. Sava, T. Giraldi, L. Lassiani, and C. Nisi, Metabolism and mechanism of the antileukemic action of isomeric aryldimethyltriazenes, Cancer Treat. Rep. 66: 1751 (1982).

12. G. Sava, S. Zorzet, L. Perissin, T. Giraldi and L. Lassiani, Effects of an inducer and an inhibitor of hepatic metabolism on the antitumor action of dimethyltriazenes Cancer Chemother.Pharmacol. 21: 241 (1988).

13. M.F.G. Stevens, J.A. Hickman, R. Stone, N.W. Gibson, E. Lunt, C.G. Newton, and G.U. Baigh, Antitumor imidazotetrazines. 1. Synthesis and chemistry of 8-carbamoyl-3-(2-chloroethyl)imidazo(5,1-*d*)- 1,2,3,5-tetrazin-4(3H)-one, a novel broad-spectrum antitumour agent,J. Med. Chem. 27: 196 (1984).

14. M.J. Tisdale, Induction of haemoglobin synthesis in the human leukemia cell line K562 by monomethyltriazenes and imidazotetrazinones, Biochem. Pharmacol. 34: 2077 (1985).

15. M. Zucchetti, C.V. Catapano, S. Filippeschi, E. Erba, and M. D'Incalci, Temozolomide induced differentiation of K562 leukemia cells is not mediated by gene hypomethylation, Biochem. Pharmacol. 38: 2069 (1989).

16. C.V. Catapano, M. Broggini, E. Erba, M. Ponti, L. Mariani, L. Citti, and M. D'Incalci, *In vitro* and *in vivo* methazolastone-induced DNA damage and repair in L1210 leukemia sensitive and resistant to chloroethylnitrosoureas, Cancer Res. 47: 4884 (1987).

17. V.L. Bull, and M.J. Tisdale,Antitumour imidazotetrazines- XVI. Macromolecular alkylation by 3-substituted imidazotetrazinones, Biochem. Pharmacol. 36: 3215 (1987).

18. A.E. Pegg, D. Scicchitano, and M.E. Dolan, Comparison of the rates of repair of O^6-alkylguanines in DNA by rat liver and bacterial O^6-alkylguanine-DNA alkyltransferase, Cancer Res. 44: 3806 (1984).

19. R.J. Graves, B.F.L. Li, and P.F. Swann, Repair of O^6-methylguanine, O^6-ethylguanine, O^6-isopropylguanine and O^4-methylthymine in synthetic oligodeoxynucleotides by Escherichia coli ada gene O6-alkylguanine-DNA-alkyltransferase,Carcinogenesis 10: 661 (1989).

20. B.W. Durkacz, O.Omidiji,D.A. Gray, and S. Shall, (ADP-ribose)$_n$ partecipates in DNA excision repair, Nature 283: 593 (1980).

21. S. Shall, ADP-ribosylation, DNA repair, cell differentiation and cancer, In: "ADP- ribosylation , DNA repair and cancer", M. Miwa et al., ed., Japan Sci.Soc.Press, Tokyo, (1983).

22. J.A. Slack, and C. Goddard, Antitumor imidazotetrazines: VII. Quantitative analysis of mitozolomide in biological fluids by high-performance liquid chromatography, J.Chromatogr. 337: 178 (1985).

CLINICAL USE OF TRIAZENES

Sándor Eckhardt

National Institute of Oncology
Budapest, Hungary

INTRODUCTION

Dacarbazin (5-methyl-(3,3-dimethyl-1-triazene,DTIC) is an alkylating agent with established clinical antitumour activity. As a single agent, DTIC explicits tumour inhibitory effect against Hodgkin's disease, melanoma and sarcomas. Certain apudomas may be also responsive to DTIC. As a results of these observations various combination regimens containing DTIC were designed and proved to be clinically effective. This paper is aimed at summarizing the clinical results of DTIC treatment so far achieved in various malignant diseases.

MALIGNANT MELANOMA

Dacarbazin was extensively studied in malignant melanoma. Single agent activity of DTIC compared to other cytostatic agents is tabulated in Table 1:

Table 1. Single agent activity in Malignant Melanoma

Drug	No of patients	CR + PR	%	Authors
DTIC	851	205	24	(1)
HU	200	24	12	(2)
VDS	165	32	19	(3)
VBL	137	28	20	(4)
CCNU	133	17	13	(1)
Me-CCNU	124	15	12	(2)
BCNU	123	22	18	(4)
MPL	110	10	9	(4)
TIC-MUSTARD	108	9	8	(1)

HU=hydroxyurea   VDS=Vindesine   VBL=Vinblastine
MPL=Melphalan

Cumulative data compiled by De Vita on 978 patients confirmed that 214 complete or partial responses (22%) were obtained with DTIC in patients with malignant melanoma (5). It is therefore obvious that among the nine drugs studied DTIC was the most promising antitumour agent suitable for combination chemotherapy studies.

Two drug combinations with DTIC are listed in the following table:

Table 2. Two Drug Combinations in Malignant
        Melanoma

Drugs	No of patients	CR+PR	%	Authors
DTIC + CPM	120	23	19	(6)
DTIC+BCNU	93	16	17	(7)
DTIC+MeCCNU	165	30	18	(8)
DTIC+VCR	20	2	10	(8)
DTIC+ACT-D	157	23	22	(9)
DTIC + PCZ	62	10	16	(10)
DTIC+VDS	14	7	50 (?)	(11)
Total	631	111	17.5	

CPM=cyclophosphamide, VCR=Vincristine, ACT-D=Actinomycin, PCZ=Procarbazin

Among two drug combinations none of the regimens studied proved to be superior to the single agent activity of DTIC. The results obtained with DTIC and Vindesin need to be confirmed.
The following three drug combinations were clinically investigated:

Table 3. Three Drug Combinations in
        Malignant Melanoma

Drugs	No of patiens	CR+PR	%	Authors
DTIC+BCNU+VCR	84	20	24	(10)
	135	31	23	(12)
	129	26	21	(13)
DTIC+BCNU+CCNU	11	4	36 (?)	(1)
DTIC+BCNU+HU	32	14	44	(12)
	89	24	27	(12)
DTIC+HU+VCR	84	16	20	(13)
DTIC+DDP+VDS	21	5	24	(14)
DTIC+CPM+ICRF-159	157	19	12	(6)
DTIC+CRM+VCR	50	20	40	(1)

DDP = Cisplatin

Results achieved with DTIC containing three drug combinations seem to be slightly superior to the single agent therapy. Especially nitrosoureas and vinca alkaloids potential candidates for effective regimens.
Four drug combinations are listed in the next table:

Table 4. Four Drug Combinations in
Malignant Melanoma

Drugs	No. of patients	CR+PR	%	Authors
DTIC+BCNU+HU+VCR	89	27	30	(1)
DTIC+BCNU+PCZ+VCR	25	7	28	(12)
DTIC+CCNU+BLM+VCR	72	29	40	(15)

At present, the combination of DTIC+CCNU+BLM+VCR (Bleo-mycin+Oncovin+Lomustin+Dacarbazin=BOLD) is the regimen most commonly used for the treatment of metastatic melanoma.

At the National Institute of Oncology in Budapest, Hungary a Mitolactol containing combination was also tested. Mitolactol (1,6-dibromo,1,6-dideoxydulcitol) is a bifunctional alkylating agent showing antitumour activity against experimental melanomas (17). Similar studies were initiated by other clinical in-vestigators as well and are enumerated in table 5:

Table 5. DBD Containing Drug Combinations in
Malignant Melanoma

Drugs	Dosage Mg/M^2/D	Days	No of patients	CR	PR	%	Authors
DTIC+ DBD	iv 800 po 125	D 1 D 1-10	14	–	4	29	(17)
⌈VBL+⌉ ⎸BLM+⎹ ⎸DDP ⌋ ⎸DTIC+⌉ ⌊DBD ⌋	iv 3 iv 15 iv 50 iv 800 po 125	D 1,2 D 1-5 D 5 D 1 D 1-10	20	–	7	35	(17)
DTIC+ DBD+ ACT-D	iv 750 po 90-190 iv 1.25	D 1 D 3-7 D 1	20	2	3	25	(18)
DTIC+ DBD+ ACT-D	iv 750 po 90-230 iv 1.25	D 1 D 3-7 D 1	37	2	9	30	(18)
DTIC+ DBD+ ACT-D	iv 750 po 150 iv 1.25	D 1 D 3-7 D 1	22	1	6	32	(19)

DBD = Dibromdulcitol (Mitolactol)

It can be stated that DTIC combined with DBD and actino-mycin-D is effective in approximately 30% of the patients. This finding has to be confirmed in a larger group of patients. Since similar results can be observed with the BOLD regimen, a randomized trial would be also requires to compare the efficacy of the two therapeutic approaches.

Despite of positive but scarce single agent data (20) DTIC was rapidly introduced into the therapy of Hodgkin's disease. Effective combination regimens were designed and some of them became standard treatment in this malignant disease.

Table 6 illustrates the DTIC containing combinations:

Table 6. DTIC Containing Drug Combination in Hodgkin's disease

Drug		Dosage	Authors
ADM ⌉		25 mg/M^2 D1, 15	
BLM	ABVD	10 V/M^2 D1, 15	(21, 22)
VBL		6 mg/M^2 D1, 15	
DTIC ⌋		375 mg/M^2 D1, 15	
ADM ⌉		45 mg/M^2 D1,	
BLM		5 mg/M^2 D1,5	
CCNU	ABDIC	50 mg/M^2 D1	(23)
DTIC		200 mg/M^2 D1-5	
PRED ⌋		40 mg/M^2 D1-5	
ADM ⌉		40 mg/M^2 Q3W	
BLM		15 U/ M^2 Q1W	
CCNU	VABCD	80 mg/M^2 Q6W	(24)
DTIC		800 mg/M^2 Q3W	
VBL ⌋		6 mg/M^2 Q3W	

Among the regimens listed above ABVD was rapidly recognized as an effective salvage therapy for MOPP or COPP resistant patients. Accordingly, it is widely used as a second-line treatment for Hodgkin's disease. On the basis of the favourable results published by several authors, the Italian group initiated a trial in which MOPP was alternated with the ABVD regimen and compared to those receiving only MOPP (25). The results can be found in the following table:

Table 7. Hodgkin's Disease: Alternating MOPP/ABVD versus MOPP.7 Years Overall Results

	MOPP	MOPP/ABVD	P-VALUE
CR	74.4	88.8	0.14
RFS	44.4	76.8	0.004
Survival (CR)	73.2	90.3	0.038
Total survival	61.1	82.3	0.043

CR = complete remission RFS = relapsefree survival

DTIC as a single agent proved to be active against soft tissue sarcomas. In a series of 53 patients Gottlieb reported a response rate of 17% (26). A literature survey performed by Rosenberg on 109 patients confirmed a 16% response rate (27). Leiomyosarcomas were more sensitive to DTIC than any other histologival classes of soft tissue malignancies. Phabdomyosarcomas and liposarcomas were less responsive, while fibrosarcomas, angiosarcomas and synovial cell sarcomas did not respond at all. Based on these data two and three drug combinations have been designed and clinically tested. The results of these combinations are shown in table 8.

Table 8. Two and Three Drug Combinations in
Soft Tissue Sarcomas.

Drug	No of patients	CR	PR	RR%	Authors
ADM+DTIC	100	7	25	32	(28)
	218	25	67	42	(29)
	79	11	14	32	(27)
	18	1	5	33	(27)
ADM+DTIC	107	10	35	42	(30)
+VCR	54	1	5	11	(31)

Table 8 clearly demonstrates the efficacy of these combinations. Approximately one third of the patients were responsive to the therapy. These encouraging results prompted the investigators to warrant four drug combinations as well. This effort resulted in the CYVADIC and CYVADACT regimens which are novadays considered as standard chemotherapeutic treatments for advanced soft tissue sarcomas. In CYVADIC DTIC and in CYVADACT Actinomycin D is added to ADM+CPM+VCR. The Activity of CYVADIC is as follows:

Table 9. Activity of CYVADIC Treatment in
Soft Tissue Sarcomas

Drug	No of patients	CR	PR	RR %	Authors
CPM + VCR +	125	21	42	50	(33)
ADM + DTIC	60	14	8	37	(32)
	229	27	80	45	(34)
	95	12	21	35	(27)

Several authors claim that CYVADIC is the best DTIC containing regimen in the treatment of soft tissue sarcomas. 35-50% response rates clearly show its effectiveness. Results obtained with CYVADACT do not differ significantly from those achieved with CYVADIC.

Uterine sarcomas were studied separately from soft tissue sarcomas. Table 10 lists data on the activity of DTIC in this malignancy:

Table 10. Activity of DTIC in Uterine Sarcomas

Drugs	No of patients	CR	PR	RR%	Authors
DTIC	24	-	6	25	(35)
DTIC+ADM	70	7	10	243	(36)
DTIC+ADM+VCR	6	3	-	50	(37)

The highest response rates were obtained in patients with heterologous mesodermal histology. Among metastatic sites the lungs were the most responsive to the treatment. Accumulation of more data with the three and four drug regimens is required. Nevertheless, it is obvious that metastatic uterine sarcomas are potential targets for DTIC containing chemotherapy.

PEDIATRIC MALIGNANCIES

DTIC containing regimens proved to be effective in various pediatric malignancies. The combinations of DTIC+CPM+ADM+VCR+ACT-D induced 87% responses in 44 children with Ewing-sarcoma (38). Similarly osteogenic sarcomas proved to be sensitive to DTIC containing combinations. Three and four drug combinations were effective in 24-35% of the cases, while the five drug combination reported above resulted in a 60% response rate (39,40)

Children's neuroblastoma was another important target of clinical investigations. Results obtained with DTIC containing combinations are enumerated in the following table:

Table 11. Activity of DTIC Containing
Regimens in Neuroblastoma

Drugs	No of patients	CR	PR	RR%	Authors
DTIC	37	1	4	14	(41)
DTIC+ADM	45	1	11	27	(42)
DTIC+VCR	28	-	3	11	(42)
DTIC+VCR	60	23	25	80	(43)
DTIC+VCR+CPM+ADM	44	19	17	81	(43)
DTIC+VCR+HN$_2$+ADM	12	2	6	66	(44)

HN$_2$= Nitrogen mustard

Despite some controversial data due to different schedules it can be stated that DTIC is an useful agent in the therapy of neuroblastoma and further trials are warranted in order to find the best combination and therapeutic schedule.

MISCELLANEOUS NEOPLASMS

The antitumour activity of DTIC was investigated in various solid tumours as well. Marginal activity was only seen in cervical cancer (45). One study reported some activity in small cell lung cancer if combined with other drugs and radiotherapy (46).

These findings need to be confirmed.
Nevertheless, another field of potential clinical application
for DTIC could be APUD-omas. According to several authors
antitumour effect was seen in malignant carcinoids (47, 48)
and hormone producing islet-cell tumours (insulimoma, glucagonoma) (49, 50). Despite the rarity of these tumours data of
several dozens of patients were compiled and approximately
80% response rates recorded.
No responses were obtained in patients with breast or colorectal cancer.

SUMMARY

The single agent activity of DTIC has been widely accepted
in malignant melanoma and the drug is successfully combined
with nitrosoureas and vinca alkaloids. The combination with
investigational drugs such as mitolactol (dibromodulcitol)
seems to be also promising. More recently its application in
chemo-immunotherapy is explored. Especially sequential treatment with Interleukin-2 is under study.
In combination with other cytostatic agents the administration of DTIC is particularly useful in Hodgkin's disease. The
combination regimen ABVD can be considered as a standard
second line therapy for MOPP or C-MOPP resistant cases. More
recently ABVD as first line therapy is tested and the sequential alteration of ABVD and MOPP is also under investigation.
Other less frequently used regimens are ABDIC or VABCD. Their
value have to be confirmed versus the standard ABVD regimen.

DTIC is also active against soft tissue sarcomas. Combinations like CYVADIC or CYVADACT are the most commonly used
regimens for the therapy of these malignancies. Among mesen-
chymal neoplasms uterine sarcoma seems to be particularly
sensitive to DTIC containing regimens.

DTIC is active in pediatric malignancies. Neuroblastomas,
Ewing sarcomas are responsive to the treatment.

DTIC was also reported to be active against some of the
apudomas.

In conclusion, DTIC has its established place in the
chemotherapy of malignant diseases.

REFERENCES

1. R.L.Comis, DTIC (NSC-45388) in Malignant Melanoma: A
   Perspective. Cancer Treatment Rep.60: 165 (1976).
2. T.H.Wassermann, M.Slavik, S.K.Carter, Clinical comparison
   of the Nitrosoureas. Cancer 36: 1258 (1975)
3. H. O.Peters, J.Becker, V.Lins, D.Preding, Eldisine (vindesin)
   in der Chemotherapie solider Tumoren und Hemoblastosen, in:
   "Spezielle Tumorchemotherapie", S.Tanneberger, et.,
   Akademie Verlag, Berlin, 1986, 334.
4. J.K.Luce, Chemotherapy and Malignant Melanoma. Cancer 30:
   16o4 (1974)
5. M.J.Mastrangelo, A.R.Baker, H.R.Katz, Cutaneous Melanoma
   In: V.T.De Vita, S.Hellman, S.A.Rosenberg,"Cancer, Principles and Practice of Oncology". J.P.Lippincott Company,
   Philadelphia, 1985. 1371.

6. C.A.Presant, A.Bartolucci, C.Balch, M.Tronier, A Randomized Comparison of Cyclophosphamide, DTIC with or without Piperazinedione in Metastatic Malignant Melanoma. Cancer 49: 1355 (1982).
7. P.P.Carbone, W.Costello, Eastern Cooperative Group Studies with DTIC (NSC-45388). Cancer Treatment Rep. 60: 193 (1976).
8. D.L.Ahmann, H.F.Bisel, J.H.Edmonson, R.G.Hahn, R.T.Eagan, M.J. O'Connel, S.Frytak, Clinical Comparison of Adriamycin and a Combination of methyl-CCNU and Imidazole Carboxamide in Disseminated Malignant Melanoma. Clin.Pharmacol.Therap. 19: 821 (1976).
9. R.E.Gerner, G.E.Moore, M.S.Didolkar, Chemotherapy of Disseminated Malignant Melanoma with Dimethyl Triazeno Imidazole Carboxamide and Dactinomycin Cancer 32: 755 (1973).
10. L.H.Einhorn, M.A. Burgess, C. Vallejos, P.Bodey, J.V.Guttermann, G.Mavligit, E.M.Hersh J.K.Luce, E.Frei, E.J.Freireich, J.A.Gottlieb, Prognostic Correlations and Response to Treatment in Advanced Metastatic Malignant Melanoma, Cancer Res. 34: 1995 (1974).
11. B.S.Yap, M.A.Burgess, R.S.Benjamin, E.M.Hersh, G.P.Bodey, DTIC and Continuous 5-day Infusion of Vindesine+i.v.MER for Metastatic Melanoma - ASCO abstracts 1: 178 (1982).
12. J.J.Costanzi, DTIC (NSC-45388) studies in Southwest Oncology Group. Cancer Treatment Rep., 60: 189 (1976).
13. G.Beretta, G.Bonadonna, N.Cascinelli, Comparative Evaluation of Three Combination Regimes for Advanced Malignant Melanoma: Results of an International Cooperative Study. Cancer Treatment Rep., 60: 33 (1976).
14. M.Graubner, G.Staebe, J.Illig, E.Paul, M.Hundeiker, H. Pralle, Cisplatin-Vindesine-Dacarbazine in Advanced Melanoma: Response Related to Site of Metastases. J.Cancer Res. Clin.Oncol. 107: 62 (1984).
15. H.F.Seigler, V.S.Lucas, N.J.Pikkett, A.T.Huang, DTIC, CCNU, Bleomycin and Vincristine (BOLD) in Metastatic Melanoma. Cancer 46: 2346 (1980).
16. S.Somfai-Relle, L.Németh, Antitumour Effect of DBD in Vivo. In:"Dibromodulcitol",S.Eckhardt ed., Medicina, Budapest, 1982, 73
17. R.P.Richman, T.H.Woodcock, T.T.Kubota, M.S.Blumenreich, P.S.Gentile, J.C.Allegra, Phase II Trial of Vinblastine, Bleomycin and Cisplatin (VBP) Followed by Dacarbazine and Mitolactol in Metastatic Melanoma. Cancer Treat.Rep. 68: 1395 (1984).
18. M.K.Samson, C.D.Haas, L.H.Baker, G.Cummings, Dibromodulcitol (DBD), DTIC and Actinomycin-D in Disseminated Malignant Melanoma. A Phase I-II Clinical Trial. Am.J. Clin.Oncol. 8: 167 (1985).
19. A.Kiskőszegi, DBD Containing combinations in Melanoma. In: Institóris L., Kaczmarek J., Jeney, A.,"Mitolactol" Bibliotheca Chinoina, Budapest, Hungary, 1986, 84.
20. E.Frei, III: Status and Perspectives in Chemotherapy of Hodgkin's Disease. Arch.Int.Med. 131: 439 (1973).
21. G.Bonadonna, R.Zucali, S.Monfardini, M.Delena, C.Uslenghi, Combination Chemotherapy of Hodgkin's Disease with Adriamycin, Bleomycin, Vinblastine and Imidazole Carboxamide Versus MOPP, Cancer 35: 252, 1975.
22. A.Santoro, V.Bonfante, G.Bonadonna: Salvage Chemotherapy with ABVD in MOPP-resistant Hodgkin's Disease. Ann. Intern.Med. 96 : 139, 1982.

23. N.Taunir, F.Hagemeister, W.Velasquez, Long-term Follow-up with ABDIC Salvage Chemotherapy of MOPP Resistant Hodgkin's Disease. J.Clin.Oncol. 1 : 432 (1983).
24. L.H.Einhorn, S.D. Williams, E.E.Stevens: The Treatment of MOPP-Refractory Hodgkin's Disease with Vinblastine, Doxorubicin, Bleomycin, CCNU and Dacarbazine, Cancer 51: 1348 (1983).
25. G.Bonadonna, P.Valagussa, A.Santoro, Alternating Non-cross Resistant Combination Chemotherapy or MOPP in Stage IV Hodgkin's Disease. A report of 8-year Results. Ann.Intern. Med. 104 : 739 (1986).
26. J.E.Gottlieb, R.S.Benjamin, L.H.Baker, Role of DTIC (NSC-45388) in the Chemotherapy of Sarcomas. Cancer Treatm.Rep. 60 : 199 (1976).
27. S.A.Rosenberg, H.D.Suit, L.H.Baker, Sarcomas of Soft Tissues In: V.T.DeVita, S.Hellman, S.Rosenberg, "Cancer Principles and Practice of Oncology", Lippincott, Philadelphia, 1985, 1281.
28. E.C.Borden, D.Amato, H.T.Enterline, H.Lerner, P.P.Carbone, Randomized Comparison of Adriamycin Regimens for Treatment of Metastatic Soft Tissue Sarcomas. Proc.AACR and ASCO, C-902, 1983.
29. J.E.Gottlieb, L.H.Baker, J.M.Quagliana, Chemotherapy of Sarcomas with a Combination of Adriamycin and DTIC. Cancer 30 : 1632, (1972).
30. J.E.Gottlieb, L.H.Baker, M.A.Burgess, Sarcoma Chemotherapy. In: Cancer Chemotherapy. Fundamental Concepts and Recent Advances. 19th Ann.Clin.Conf. in Cancer, 1974, Year Book Medical Publishers. 1975.
31. E.T.Creagan, R.G.Hahn, D.L.Ahmann, A Comparative Clinical Trial Evaluating the Combination of Adriamycin, DTIC and Vinblastine, the Combination of Actinomycin D and Cyclophosphamide for advanced sarcoma. Cancer Treatm.Rep. 60: 1385 (1976).
32. H.M.Pinedo, C.P.J.Vendrik, V.H.C.Bramwell, Reevaluation of the CYVADIC Regimen for Metastatic Soft Tissue Sarcoma, Pro.AACR and ASCO C-228, 1979.
33. B.Yap, L.H.Baker, J.G.Sinkovics, Cyclophosphamide, Vincristine, Adriamycin and DTIC (CYVADIC) Combination Chemotherapy for the Treatment of Advanced Sarcomas. Cancer Treatm. Rep. 64 : 93 (1980).
34. R.S.Benjamin, J.A.Gottlieb, L.H.Baker, CYVADIC Versus CYVADACT. A Randomized Trial of Cyclophosphamide (CY), Vincristine (V) and Adriamycin (A) plus either Dacarbazin (DIC) or Actinomycin D (DACT) in Metastatic Sarcomas. Proc.AACR and ASCO C-256, 1976.
35. H.Pinedo, Y.Kenis, Chemotherapy of Advanced Soft Tissue Sarcomas in Adults. Cancer Treat.Rev. 4: 67 (1977).
36. G.A.Omura, F.J.Major, J.A.Blessing, T.V.Sedlacek, J.T. Thigpen, W.T.Creasman, R.J.Zino, A randomized study of Adriamycin with or without Dimethyltriazenoimidazole Carboxamide in Advanced Uterine Sarcomas. Cancer 52 : 626 (1983)
37. F.Aziza, J.Bitran, G.Javehari, A.L.Herbst, Remission of Uterine Leiomyosarcomas Treated with Vincristine, Adriamycin and Dimethytriazenoimidazolecarboxamide. Ann.J. Obstet.Gynecol. 133 : 379 (1979).
38. T.Vietti, Metastatic Ewing, Med.Pediat.Oncol. 7 : 61 (1979)

39. W.W.Sutow, Multidrug Chemotherapy in Osteosarcoma.<u>Clin.
    Orthop.</u> 153 : 67 (1980)
40. W.W.Sutow, M.P.Sullivan, D.J.Fernbach, Adjuvant Chemo-
    therapy in Primary Treatment of Osteogenic Sarcoma.
    <u>Cancer</u> 36 : 1598 (1975)
41. M.Carli, A.A.Green, F.A.Hayes, Therapeutic Efficacy of
    Single Drugs for Childhood Neuroblastoma: A review <u>In:</u>
    "Pediatric Oncology"Excerpta Medica, Amsterdam, Oxford,
    Princeton, 1982, 141.
42. M.Carli, G.Pastore, G.Perilongo, L.Zanesco, The Role of
    Chemotherapy in Neuroblastoma. <u>In:</u> "Pediatric Oncology",
    Excerpta Medica, Amsterdam, Oxford, Princeton, 1982, 151.
43. J.Z.Finklestein, M.F.Klemperer, A.Evans: Multiagent Chemo-
    therapy for Children with metastatic Neuroblastoma. A
    report from CCSG <u>Med.Pediatr.Oncol.</u> 6 : 179 (1979).
44. D.Traggis, N.Jaffe, D.Nathan, Advanced Neuroblastoma;
    Treatment with Combination Chemotherapy: Vincristine,
    Adriamycin, Nitrogen Mustard and DTIC (VAM-DTIC) <u>Proc.</u>
    <u>Amer.Ass.Cancer Res.</u> 17 : 140 (1976).
45. T.H.Wassermann, R.L.Comis, M.Goldsmith, H.Handelsmann,
    J.S.Penta, M.Slavik, W.T.Soper, S.K.Carter, Tabular
    Analysis of the Clinical Chemotherapy of Solid Tumours.
    <u>Cancer Chemother. Rep.</u> 6 : 399 (1975).
46. N.J.Perevodcsikova, M.B. Büchkov, G.A.Tevzadze, Intensive
    Combination Chemotherapy combined with radiotherapy in
    the Treatment of Small Cell Lung Cancer (in Russian).
    <u>Vestnik AMN SSSR</u> 7 : 60 (1981).
47. C.G.Moertel, Treatment of the Carcinoid tumour and the
    Malignant Carcinoid Syndrome. <u>J. Clin. Oncol.</u> 1 : 727
    (1983).
48. A.Kessinger, J.F.Foley, H.M.Lemon, Use of DTIC in the
    Malignant Carcinoid Syndrome. <u>Cancer Treatm. Rep.</u> 61 :lol
    (1977).
49. J.F.Foley, H.M.Lemon, DTIC Therapy for Malignant Islet
    Cell Tumours. <u>Proc. ASCO</u> 1 : 169 (1982)
50. A.Kessinger, H.M.Lemon, J.F.Foley, The Glucagonoma and
    its Management. <u>J. Surg. Oncol.</u> 9 : 419 (1977).

CLINICAL STUDIES WITH THE p-CARBOXYL DIMETHYL PHENYL TRIAZENE CB10-277

David R Newell[1], Brenda J Foster[1], James Carmicheal[2], Adrian L Harris[2] Karen Jenns[1], Lyndsey Gumbrell[1] and A Hilary Calvert[1]

1. Institute of Cancer Research, Sutton, Surrey, England
2. University of Newcastle upon Tyne, Newcastle, England

INTRODUCTION

Analogue development is one of the undoubted success stories of cancer chemotherapy. For example, there are many analogues of the original nitrogen mustard HN2; methotrexate is an analogue of the earlier drug aminopterin; epirubicin is a less cardiotoxic analogue of doxorubicin and carboplatin is a platinum complex which, unlike cisplatin, lacks significant nephrotoxicity. Although DTIC (dacarbazine) does not come into the same class as the above groups of compounds, either in terms of its spectrum or level of activity, it is deemed to be a useful drug. Thus DTIC remains one of the most active single agents for the treatment of malignant melanoma. It is also used in the management of lymphoma and sarcoma although, since DTIC is part of drug combinations in these latter diseases, its exact contribution to therapeutic success is difficult to evaluate.

Although it has been in use for 20 years there remain a number of unresolved issues with respect to the clinical application of DTIC. It has been known for some time that DTIC can undergo extensive metabolism and this has been implicated as being an important requirement for antitumour activity (studies reviewed in (1) and by A Gescher in this volume). Briefly, DTIC metabolism involves oxidative N-demethylation to form a monomethyl metabolite which in turn can methylate cellular macromolecules. The formation of the monomethyl metabolite is thought by many to be a critical step in determining the activity of the drug. Preliminary studies ((2) and C J Rutty personal communication) suggest that in patients the formation of the monomethyl metabolite of DTIC may not be sufficient to allow the full therapeutic potential of the drug to be revealed and further studies are required to address this point. The second area of controversy in the clinical use of DTIC is the role of photo decomposition as a determinant of the efficacy of the drug. Although DTIC is photo labile (3-5), and it has been suggested that guarding against this is advantageous (6), there are no controlled clinical studies to prove that protection of DTIC solutions from light is necessary. Indeed, in one recent carefully conducted preclinical study exposure of DTIC injection solutions to light prior to adminstration had no statistically significant effect on *in vivo* antitumour activity (7). If anything, there was a trend towards greater activity with drug that had been exposed to light (7). However these latter authors did observe that mice treated with DTIC

Triazenes, Edited by T. Giraldi *et al.*
Plenum Press, New York, 1990

intradermally and then exposed to light did suffer skin ulceration. Although this is not a major clinical problem with DTIC there have been case reports of photosensitivity reactions (8). Thus there is no clear consensus in the literature as to the extent to which DTIC photo lability is a problem. The third area in which the clinical use of DTIC might be improved relates to drug dose and scheduling. DTIC is usually given on a daily x5 schedule to reduce the incidence of nausea and vomiting. However recent studies have shown that single doses as high as $1.5g/m^2$ can be given and that at this dose the limiting toxicity is haematological (9). Again clinical trials are required to define the most therapeutic schedule. A final and relatively minor concern in the clinical use of DTIC is that, due to the insolubility of the drug at neutral pH, it is administered in an acidic injection vehicle (ca. pH4) and it is not known whether or not this practise is devoid of problems.

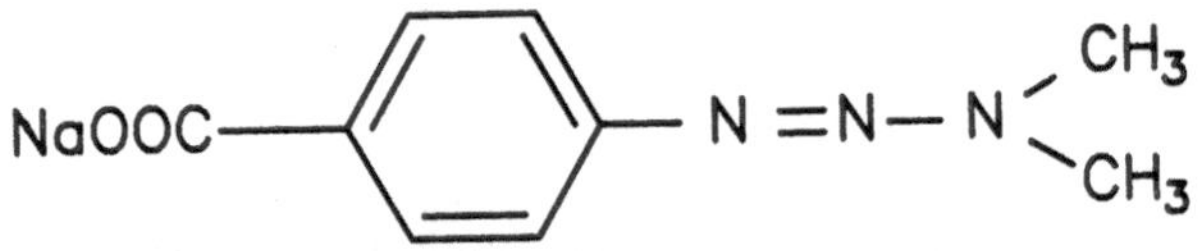

Figure 1.   The structure of CB10-277

In view of the above uncertainties over the clinical use of DTIC an analogue programme was initiated at the Institute of Cancer Research which, as information on DTIC accrued, developed with two broad aims. Firstly, a DTIC analogue was sought which was stable to light and adequately soluble as neutral pH and, secondly, the candidate drug was required to demonstrate satisfactory *in vivo* metabolic activation. Additional requirements were that the analogue should display equivalent or superior activity to DTIC in experimental antitumour models and that the dose limiting toxicity in preclinical models should be clinically acceptable. In order to achieve the required stability the phenyl ring was chosen as a substitute for the imidazole moiety of DTIC and to achieve adequate aqueous solubility the carboxylic acid derivative (benzoic acid) was chosen. Against the TLX5 lymphoma the isomer with the carboxylic acid *para* to the triazeno function had the highest therapeutic index (10), a result which was confirmed independently and extended to the P388 leukemia (11). The compound with a dimethyl triazene function, ie. directly analogous to DTIC, was amongst the compounds with the highest level of activity against the TLX5 lymphoma (12) and hence this compound (CB10-277, DM-COOK (potassium salt), figure 1) was selected for detailed preclinical investigation.

CB10-277 has been independently synthesised and studied by a number of groups such that there is now a large body of preclinical information available on the compound. CB10-277 has *in vivo* activity against a broad range of experimental tumours both rodent and human in origin (for example, (10-14) and B J Foster, D

R Newell and M Jones unpublished results). Where comparative studies have been performed, the level and spectrum of activity of CB10-277 is broadly similar to that of DTIC. To evaluate whether or not CB10-277 was subject to satisfactory metabolic activation *in vivo*, studies were performed in mice with analysis of parent compound and monomethyl metabolite levels by HPLC. Following the adminstration of an LD10 dose of CB10-277 the monomethyl metabolite was readily detected in the plasma of mice at levels at which monomethyl triazenes are known to be cytotoxic in *in vitro* (15,16). In addition to CB10-277 and the monomethyl metabolite two conjugates of the parent compound were detected, namely, the glycine and glucuronic acid conjugates. The data for the pharmacokinetics of CB10-277 in the mouse are summarised in table 1 and the proposed metabolic pathway for CB10-277 given in figure 2.

As a final prelude to the clinical evaluation of CB10-277 the toxicology of the drug was investigated by the British Industrial Biological Research Association under the auspices of the Cancer Research Campaign (UK) Clinical Trials Committee. These studies showed that following a single dose the LD10 values for iv and ip administration were 265(218-323) mg/kg and 308(235-404) mg/kg, respectively. The only major adverse effects were haematological as shown by anaemia, leucopenia and thrombocytopenia. In addition there was bone marrow, thymus, spleen, lymph node and testicular atrophy. Studies of CB10-277 given by repeated ip administration to either mice or rats failed to reveal any other major toxicities.

Table 1. Summary of the pharmacokinetics of CB10-277 in Balb C mice (250mg/kg iv)

Compound	CB10-277	Monomethyl Metabolite	Glycine Conjugate	Glucuronide Conjugate
Peak plasma level (uM)	ca.2000	97 ± 23	ND	1021 ± 106
Plasma AUC (mMxmin)	142	8	ND	89
Half life (min)	32 ± 0.4	32 ± 2.4	ND	26 ± 2.1
24h Urinary excretion (% dose admin)	0.3 ± 0.3	ND	0.4 ± 0.1	38 ± 9

ND = Not detected; peak plasma level is the mean $\pm$ SD of 4 determinations; AUC is the area under the plasma concentration v time curve; half lives are the mean $\pm$ SE from the non-linear least squares analysis of plasma concentration v time data and 24h excretion is the mean $\pm$ SD from 10 individual mice. All results are taken from (14).

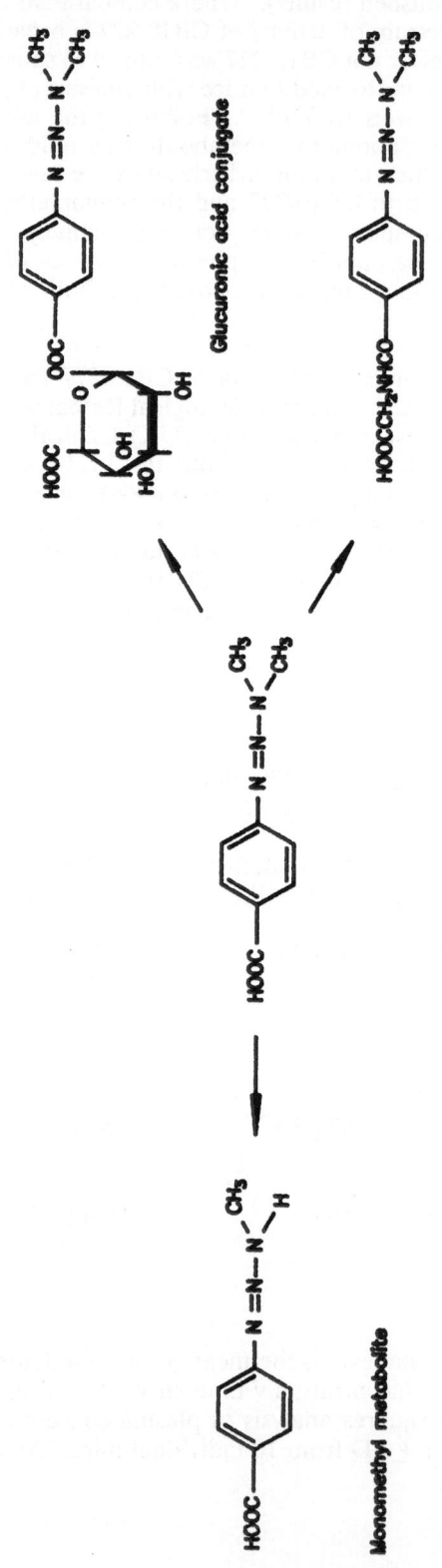

Figure 2.   The proposed metabolic pathway for CB10-277

Thus CB10-277 was selected for clinical evaluation having satisfied the criteria defined for the identification of a second generation dimethyltriazene. CB10-277 displayed satisfactory photo stability and solubility ((12) and BJ Foster and C J Rutty unpublished results), the drug had equivalent antitumour activity to DTIC in experimental systems and the dose limiting toxicity was clinically acceptable, ie. haematological. In line with current practice the phase I starting dose for CB10-277 ($80mg/m^2$) was 1/10th the mouse LD10 ($265mg/kg$ = ca. $800mg/m^2$) when the drug was given by the route and schedule intended for the clinical study. The single bolus dose administration schedule was chosen as this is the most practicable clinical schedule and there are no preclinical data to suggest that repeat administration is required for activity. Pharmacokinetic studies were performed as an integral part of the evaluation of CB10-277 in order to monitor levels of the both the parent compound and monomethyl metabolite and thereby provide information which might aid dose escalation. Finally, patient accrual was biased towards patients with DTIC sensitive tumour types as preclinical studies indicated a similar spectrum of activity for the two drugs and hence these patients were those most likely to benefit from therapy with a DTIC analogue.

Certain aspects of the data presented in this article have been given previously in abstract form or will be the subject of future full publications.

METHODS

Compounds and chemicals

CB10-277 (sodium salt, MW 215) was supplied as a lyophilised, pyrogen and preservative-free powder by the Developmental Therapeutics Programme, National Cancer Institute, Bethesda, MD, USA. The monomethyl metabolite of CB10-277 was synthesised as the potassium salt by Professor C Nisi, University of Trieste, Trieste, Italy and provided as a generous gift by Dr M D'Incalci, Mario Negri Institute, Milano, Italy. The glycine conjugate of CB10-277 was synthesised by Dr DEV Wilman, Institute of Cancer Research, Sutton, Surrey. All other chemicals were obtained from standard suppliers and were analytical grade where available.

Patient treatment and pharmacokinetic sample collection

The full details of the phase I clinical investigation of CB10-277 will be published elsewhere ((14) and BJ Foster et al. manuscript in preparation). All patients had metastatic disease either refractory to standard treatment or for which no conventional therapy exists. Performance status of less than or equal to WHO 2 was required as was satisfactory pretreatment haematological status and liver and renal function. Response and toxicity were graded by standard WHO criteria. Informed consent was obtained following the guidelines of the local Ethical Committees and the Royal College of Physicians, London. The phase I evaluation of CB10-277 was carried out under the auspices of the Cancer Research Campaign (UK) Clinical Trials Committee. The characteristics of the patients studied are given in table 2.

The CB10-277 starting dose was $80mg/m^2$ which was given as an iv bolus dose at a concentration of $50mg/ml$ in $0.9\%NaCl$, this concentration was kept constant throughout the short infusion study. Eleven escalation steps were used to reach the maximum tolerated dose of CB10-277 given by short infusion with the administration time within 5-20min for all dose levels. Thereafter further dose escalation was acheived by giving CB10-277 as a 24h infusion when the drug was dissolved in 2l

0.9%NaCl and given was the aid of an infusion pump. Patients who received more than one course of treatment were retreated every 21 days. Blood samples were collected from an indwelling intravenous canula placed in the arm opposite to that receiving the drug. Plasma samples were prepared from chilled (0°C) heparinised blood immediately after collection and stored at -20°C until analysed. Typically, blood samples were taken 5, 10, 15 ,30, 60, 120, 240, 480, 720, and 1080 min after the finish of the short infusion of CB10-277. For patients treated with CB10-277 by 24h infusion blood samples were taken 60, 120, 240, 360, 720, 1440, 1445, 1450, 1455, 1470, 1560 and 1680 min after the start of the infusion.

Pharmacokinetic studies

Plasma levels of CB10-277, the monomethyl metabolite and the glycine and glucuronide metabolites were measured by reverse phase HPLC. Prior to analysis plasma samples were thawed on ice and plasma proteins precipitated by the addition of 2 volumes of ice cold methanol followed by centrifugation (1500g, 0-4°C). Aliquots of the methanolic supernatants were kept cold (<4°C) prior to analysis on a 15x0.46cm Spherisorb C8 column fitted with a CO:PELL ODS precolumn. The column was eluted isocratically at 1.5ml/min with 15:85 methanol:0.05M ammonium acetate (v:v) and CB10-277 and its metabolites detected in the column effluent by UV absorbance at 280nm and 313nm. In this system the order of elution was monomethyl metabolite, CB10-277, glycine conjugate and glucuronide conjugate with retention times of approximately 3, 8, 11 and 14 min, respectively. Quantitation was by external standardization using synthetic CB10-277, monomethyl metabolite and glycine conjugate prepared in control human plasma at a concentration of 20uM. CB10-277 glucuronide was quantitated by using the peak area/concentration ratio for CB10-277 since it was found that the glucuronide could be quantitatively converted by beta-glucuronidase treatment to the parent compound. The HPLC assay for CB10-277 and its metabolites was linear over the concentration range 1-500uM (1-5000uM for CB10-277) with a limit of detection of 1uM. Intra and inter assay coefficients of variation were <20% on all occasions and <10% on most. Compounds present in the HPLC effluent were identified by their retention volumes and 313/280nm absorbance ratios relative to synthetic compounds prepared in human plasma and analysed in the same HPLC run. Since the synthetic glucuronide of CB10-277 was not available this compound was identified as a beta-glucuronidase sensitive component (2 sources of enzyme used with the reaction shown to be inhibited by saccharic acid 1,4-lactone) which on enzymic hydrolysis gave a component which was chromatographically indistinguishable from CB10-277. Recently, the structure of the glucuronide metabolite, extracted from human urine, has been confirmed by mass spectrometry (G Poon, personal communication).

The CB10-277 plasma concentration/time profile was studied by non-linear least squares analysis (17) using either a mono or biexponential equation for the short infusion study with correction being made for infusion times of >10min (18). The CB10-277 AUC was calculated by integration of the fitted equation and the half life determined from the first order rate constants. The AUC values for the metabolites and for CB10-277 given by 24h infusion were calculated by the trapezoidal rule.

RESULTS

CB10-277 administered by short infusion

The starting dose for the phase I evaluation of CB10-277 was 80mg/m². Subsequent

Table 2. Characteristics of patients treated in the phase I evaluation of CB10-277

Schedule		Short infusion	24h infusion
Total patient number		36	22
Total number of courses		80	42
Patients lost to follow-up		4	3
Female		12	8
Male		24	14
Median age		44	55
Diagnosis:	Melanoma	11	8
	Sarcoma	7	8
	Other	18	6
Performance status:	0-1	19	6
	2	13	15
	3	4	1

dose escalation steps were 160, 240-264, 400, 600, 900, 1350, 2000, 2400, 2800, 3600, 4700 and 6000 mg/m^2. The major dose limiting toxicity encountered was nausea and vomiting the incidence of which is given in table 3. Despite the use of standard antiemetics, the nausea and vomiting was deemed to be unacceptable at 6000mg/m^2 and hence further dose escalation with the short infusion schedule was not attempted. Other side effects observed were a flushing or warm sensation in nearly all patients at doses of >1350mg/m^2, diarrhoea (7/76 courses WHO grade 1) and occasional (<5/76 courses) perspiration, altered taste, abdominal pain, malaise, visual changes and rash. All of the minor side effects were reversible and of no major clinical significance. In no patient was there any evidence of haematological toxicity.

Table 3. The incidence of nausea and vomiting in patients treated with CB10-277 by short infusion

Dose level (mg/m^2)	Courses evaluated	Number of courses with nausea and vomiting (WHO grade)			
		1	2	3	4
<900	15	-	-	-	-
900	5	1	2	-	-
1350	6	-	2	3	-
2000	9	-	-	6	-
2800	10	1	1	5	-
3600	13	-	-	11	1
4700	7	-	-	7	-
6000	7	-	-	7	-

Pharmacokinetic studies were performed after 42 courses of CB10-277 administered by short iv infusion. There was a linear increase in CB10-277 AUC with dose (r=0.92) and the plasma AUC of CB10-277 consistently exceeded that seen in mice at an LD10 dose (142 mMxmin) in all patients treated at 2000mg/m^2 and above. The glycine and the glucuronide conjugates of CB10-277 were detected in the plasma of patients with the AUC values for the latter also increasing linearly with dose (r=0.93). However the increase in the AUC of the glycine conjugate was only linear up to 900mg/m^2 CB10-277 after which no further clear increase in the AUC of the glycine conjugate was observed thereby indicating saturation of the formation of this metabolite. The glucuronide was thus the major plasma metabolite of CB10-277 detected in patients at all except the very lowest dose levels.

Despite achieving plasma levels of CB10-277 similar to those seen in mice at an LD10 dose, the monomethyl metabolite was not convincingly detected until the 3600mg/m^2 CB10-277 dose level at which point the CB10-277 AUC was equal to or greater than 300 mMxmin, ie. approximately twice that seen in mice. Even then the levels of the metabolite in the three patients studied (peak levels 5,11 and 16 uM; AUC 1, 1, and 1 mMxmin) were approximately 1/10th those seen in mice. Further dose escalation of CB10-277 given by short adminstration failed to substantially improve on this. For example at 6000 mg/m^2 CB10-277 in the two patients studied the peak monomethyl levels were 18 and 30 uM and the AUC values 2 and 4 mMxmin. Since these metabolite levels were only achieved in the face of considerable toxicity, further dose escalation with the short infusion schedule was not attempted. Instead, a 24h infusion of CB10-277 was investigated.

CB10-277 administered by 24h infusion

The starting dose for the evaluation of CB10-277 when given by 24h infusion was 4700mg/m^2, ie. one dose level below that reached when the drug was given by short infusion. Dose levels of 6000, 8000, 12000 and 15000 mg/m^2 were investigated. As with the short infusion, nausea and vomiting was the most frequent side effect. Since all patients received routine antiemetics it was not possible to use the WHO grading scheme hence the grading scheme defined in table 4 was developed. As can be seen most patients suffered nausea and vomiting which lasted up to 26h but no patient suffered intractable emesis. The only other toxicity seen with any frequency was diarrhoea which occurred in 13 of 31 evaluated courses and was WHO grade 1 or 2 in most cases. There were also isolated case of hallucinations, malaise, muscle ache, headache and flushing. Probably the most significant finding was that at a dose of 12000 mg/m^2 there was clear evidence of haematological toxicity. This was in contrast to lower doses (4700-8000mg/m^2) where in 31 evaluable courses no evidence of haematological toxicity was seen. Thus following 4 of 8 courses at 12000 mg/m^2 there was both leucopenia (1/4 grade 1, 2/4 grade 2 and 1/4 grade 4) and thrombocytopenia (3/4 grade 2 and 1/4 grade 4). Both of the two patients treated at 15000 mg/m^2 died suddenly of unknown causes before follow-up was possible (days 3 and 10 post treatment) and 12000 mg/m^2 was defined as the maximum tolerated dose with haematological toxicity as the dose limiting side effect.

Pharmacokinetic studies were performed after 12 courses of CB10-277 given as a 24h infusion. The AUC of CB10-277 again increased with dose and the glucuronide conjugate was the major plasma metabolite with the glycine conjugate also being detected, albeit at much lower levels (3-80 fold depending on dose). As had been seen at the higher dose levels of the short infusion, there was no clear evidence of a dose dependent increase in the AUC for the glycine conjugate

Table 4. The incidence of nausea and vomiting in patients treated with CB10-277 by 24h infusion

Dose level (mg/m²)	Courses evaluated	Number of courses with nausea and vomiting (grade)			
		1	2	3	4
4700	9	-	2	7	-
6000	12	-	-	12	-
8000	5	-	2	3	-
12000	8	1	1	4	-
15000	2	-	-	2	-

Grading for nausea and vomiting: Grade 1 - nausea only, grade 2 - nausea and vomiting for up to 12h, grade 3 - nausea and vomiting lasting up to 26h, grade 4 - intractable nausea and vomiting lasting for more than 26 h.

confirming that the formation of this metabolite was saturated. In contrast, there was a dose dependent increase in the plasma AUC of the glucuronide. The monomethyl metabolite was readily detected in the plasma of the majority of the patient studied (10/12 courses) with, following 12000 and 15000 mg/m², mean plasma AUC values similar to that seen in mice following an LD10 dose, ie. 8mMxmin. Thus in the 5 patients studied at 12000 and 15000 mg/m² the monomethyl metabolite AUC values were <1, 3, 8, 9 and 9 mMxmin. The achievement of these monomethyl metabolite AUC values was seen, in addition to the onset of haematological toxicity, as a justification for defining 12000 mg/m² as the recommended dose for phase II studies.

Tumour responses in patients treated with CB10-277

The response data for CB10-277 when given as a short infusion are shown in table 5. As can be seen there was clear evidence of tumour response, particulary in patients with metastatic melanoma. Of the responding melanoma patients 2 of the 4 had not received prior DTIC treatment. Following CB10-277 administration as a 24h infusion there was one mixed response of 2 months duration in a patient with recurrent melanoma previously treated with DTIC.

Table 5. Tumour responses in patients treated by short infusion during the phase I evaluation of CB10-277

Disease	Patients evaluable	Response			
		CR	PR	MR	PD
Melanoma	11	1	2	1	7
Sarcoma	7	-	-	1	6

CR - complete response, PR - partial response, MR - mixed response, PD - progressive disease.

DISCUSSION

This paper represents an overview of the recent early clinical studies with the p-carboxyl dimethyl phenyl triazene CB10-277. This agent, unlike DTIC, is readily soluble in aqueous media at neutral pH and does not appear to be subject to problems of photo decomposition. In preclinical models CB10-277 had equivalent antitumour activity to DTIC and the dose limiting toxicity was haematological. Like DTIC, CB10-277 undergoes extensive *in vivo* metabolism which includes the formation of an N-monomethyl triazene metabolite. However, unlike DTIC, CB10-277 is also subject to extensive conjugation, *via* the carboxylic acid moiety, primarily with glucuronic acid. Despite the extensive conjugation of CB10-277 in mice, the drug does give rise to levels of the monomethyl metabolite which, from *in vitro* cytotoxicity studies, would be predicted to be cytotoxic. Thus CB10-277 was selected for clinical evaluation as a DTIC analogue with improved solubility, stability and, possibly, metabolic activation.

The initial phase I study of CB10-277 utilised a short infusion bolus adminstration schedule and this showed that nausea and vomiting was the major side effect and that this was dose limiting at 6000mg/m². No haematological toxicity was seen and no other major adverse reactions observed. Pharmacokinetic studies performed in conjunction with the short infusion phase I investigation showed that levels of CB10-277 were achieved which exceeded those found in mice following the LD10 dose. Despite this, the levels of the monomethyl metabolite detected were substantially lower than in mice following an LD10 dose. In addition to N-demethylation, there was also extensive conjugation of the parent drug in patients, with the glucuronide being the major plasma metabolite of CB10-277. Lower levels of the glycine conjugate were observed and the formation of this metabolite appeared to be saturated at doses of 900 mg/m² and above. By extending the duration of infusion to 24h it was possible to escalate the dose of CB10-277 further and at a dose of 12000mg/m² clear evidence of haematological toxicity was seen which was deemed to be dose limiting. In addition, levels of the monomethyl metabolite observed at this dose level were similar to that seen in mice at the LD10 dose level where haematological toxicity was also dose limiting. The lack of haematological toxicity seen with CB10-277 at dose levels < 12000mg/m² in patients, where lower monomethyl levels were seen, makes it tempting to speculate that this side effect is a direct result of the formation of this metabolite. However, it should be pointed out that in individual patients there was not always a correlation between the formation of the monomethyl metabolite and subsequent haematological toxicity.

Although the major aim of a phase I study is not to identify antitumour activity it is always gratifying when responses are seen. Thus the observation of activity in the studies reviewed herein is very encouraging and clearly indicates that further investigations with CB10-277 are warranted. Exactly how these investigations should be performed needs to be decided in the light of current information on the mechanism of action and pharmacokinetics of dimethyltriazenes and other related compounds (see other articles in this volume). Thus the O⁶-alkylguanine alkyltransferase level in human tumours may well be a major determinant of the response of patients to CB10-277 as may be the ability of individual patients to activate the drug. Furthermore, with evaluation of the methylimidazotetrazine temozolomide, as well as the re-evaluation of high dose DTIC, phase II studies of single compounds may not be the best way to proceed. It is possible that an early comparison of the activity of different methyl triazenes maybe preferable. If such studies are supported by laboratory investigation of both pharmacokinetics and repair enzyme status the identification of the most therapeutic methyl triazene or methyl triazene prodrug should be expedited.

Returning to CB10-277, there are already a number of interesting questions raised by the use of the compound. Probably the most important relates to the role of metabolism to the monomethyl triazene as a determinant of drug activity. Studies by Sava, Giraldi and co-workers (11,19-22) have shown that CB10-277 can have an antimetastatic effect in murine metastasis models. These authors have made the important observation that CB10-277 does not readily undergo N-demethylation *in vitro*, when hepatic preparations are used, and that pretreatment with phenobarbitone does not enhance the activity of CB10-277 against the Lewis lung tumour (11,22). However, without *in vivo* pharmacokinetic experiments to determine the effect of the use xenobiotic metabolism inducers and inhibitors on the full spectrum of CB10-277 metabolism, it is difficult to interpret these data fully. The failure to detect significant N-demethylation of CB10-277 *in vitro* may be a reflection of the artificial conditions which exist in such studies or, alternatively, it may indicate that this metabolic step is being performed *in vivo* in organs other than the liver. As reported in the work reviewed herein, the monomethyl metabolite of CB10-277 is formed in both mice and patients and hence it must remain an important candidate for the species responsible for the activity of the drug although of course other drug derived materials and indeed the parent compound itself cannot be excluded.

With regard to metabolites of CB10-277 other than the monomethyl compound, the glucuronide conjugate is clearly an important species. To what extent the formation of this metabolite represents simply detoxification must await the synthesis and full biological characterisation of the molecule. This metabolite has previously been identified in mice (23) and although it is the major urinary CB10-277 metabolite it may allow the prolonged release of CB10-277 following the action of beta-glucuronidase. The synthesis of this metabolite is thus a clear priority for further work with CB10-277.

In conclusion, the work reviewed in this paper has shown that the p-carboxyl dimethyl triazene CB10-277 can be administered to patients by a schedule in which haematological toxicity is dose limiting. Under these conditions the plasma AUC of the monomethyl metabolite is similar to that seen in mice at the LD10 dose level where myelosuppression is also dose limiting. CB10-277 has shown clear evidence of activity and hence it warrants further clinical evaluation. This should be done in the light of current information on the role of metabolism and DNA repair as determinants of the activity of the dimethyl triazenes. Most probably, a randomised study against DTIC for the treatment of malignant melanoma with full laboratory support is the highest priority.

ACKNOWLEDGEMENTS

The authors are grateful to the Cancer Research Campaign UK and the North of England Cancer Research Campaign for funding the studies described in the paper. In addition they would like to thank the Developmental Therapeutics Programme of the NCI, Bethesda for the supply of the formulated CB10-277. Finally they would like to thank the medical and nursing staffs of the Royal Marsden Hospital and Newcastle General Hospital for their care of the patients who took part in these studies.

REFERENCES

1. DR Newell, A Gesher, SJ Harland, D Ross and CJ Rutty, N-methyl antitumour agents. A distinct class of anticancer drugs? Cancer Chemother Pharmacol. 19:91 (1987)

2. CJ Rutty, DR Newell, RB Vincent, G Abel, PM Goddard, SJ Harland and AH Calvert. The species dependent pharmacokinetics of DTIC. Brit J Cancer 48:140 (1983)

3. GM Baird and MLN Willoughby. Photodegradation of dacarbazine. Lancet ii:681 (1978)

4. JK Horton and MFG Stevens. Triazenes and related products. Part 23. New photoproducts from 5-diazoimidazole-4-carboxamide (diazoIC). JCS Perkins I 1433 (1981)

5. JK Horton and MFG Stevens. A new light on the photo-decomposition of the antitumour drug DTIC. J Pharm Pharmacol 33:808 (1981)

6. OM Koreich and VS Shukla. Dacarbazine (DTIC) in malignant melanoma: Reduced toxicity with protection from light. Clin Radiol 32:53 (1981)

7. RT Dorr, DS Alberts, J Einspahr, N Mason-Liddil annd M Soble. Experimental dacarbazine antitumour activity and skin toxicity in relation to light exposure and pharmacologic antidotes. Cancer Treat Rep 71:267 (1987)

8. TM Beck, NE Hart and CE Smith. Photosensitivity reaction following DTIC administration: Report of two cases. Cancer Treat Rep 64:725 (1980)

9. G Harman, D Tweedy, J Craig and D Von Hoff. Phase I-II evaluation of DTIC in the treatment of malignant melanoma on a single dose schedule. Proc 6th NCI-EORTC Symposium on new drugs in cancer therapy 418 (1989)

10. TA Connors, PM Goddard, K Merai, WCJ Ross and DEV Wilman. Tumour inhibitory triazenes: Structural requirements for an active metabolite. Biochem Pharmacol 25:241 (1976)

11. G Sava, T Giraldi, L Lassiani and C Nisi. Metabolism and mechanism of the antileukaemic action of isomeric aryldimethyltriazenes. Cancer Treat Rep 66:1751

12. DEV Wilman and PM Goddard. Tumour inhibitory triazenes. 2. Variation of antitumour activity within an homologous series. J Med Chem 23:1052 (1980)

13. T Colombo and M D'Incalci. Comparison of the antitumour activity of DTIC and 1-p-(3,3-dimethyl-1-triazeno) benzoic acid potassium salt on murine transplantable tumours and their haematological toxicity. Cancer Chemother Pharmacol 13:139

14. BJ Foster, DR Newell, J Carmicheal, AL Harris, LA Gumbrell, M Jones, PM Goddard and AH Calvert. Preclinical, phase I and pharmacokinetic studies with the dimethyl phenyltriazene CB10-277. Brit J Cancer Submitted.

15. NW Gibson, JA Hartley, RJ LaFrance and K Vaughan. Differential cytotoxicity and DNA-damaging effects produced in human cells of the Mer$^+$ and Mer$^-$ phenotypes by a series of 1-aryl-3-alkyltriazenes. Cancer Res 46:4999 (1986)

16. NW Gibson, JA Hartley, RJ LaFrance and K Vaughan. Differential cytotoxicity and DNA-damaging effects produced in human cells of the Mer$^+$ and Mer$^-$ phenotypes by a series of alkyltriazenylimidazoles. Carcinogenesis 7:259 (1986)

17. RI Jennrich and PF Sampsom. Application of stepwise regression to nonlinear least squares estimation. Technometrics 10:63 (1968)

18. JCK Loo and S Reigalman. Assessment of pharmacokinetic constants from post infusion blood curves obtained after iv infusion. J Pharm Sci 59:53 (1970)

19. G Sava, T Giraldi, L Lassiani and C Nisi. Mechanism of the antimetastatic action of the dimethyltriazenes. Cancer Treat Rep 63:93 (1979)

20. T Giraldi, G Sava, R Cuman, C Nisi and L Lassiani. Selectivity of the antimetastatic and cytotoxic effects of 1-p-(3,3-dimethyltriazeno)benzoic acid potassium salt, ($\pm$)-1,2-di(3,5-dioxopiperazin-1-yl)propane and cyclophosphamide in mice bearing Lewis lung carcinoma. Cancer Res 41:2524 (1981)

21. G Sava, T Giraldi, C Nisi and G Bertoli. Prophylactic antimetastatic treatment with aryldimethyltriazenes as adjuvants to surgical tumour removal in mice bearing Lewis lung carcinoma. Cancer Treat Rep 66:115 (1982)

22. G Sava, S Zorzet, L Perissin, T Giraldi and L Lassiani. Effects of an inducer and an inhibitor of hepatic metabolism on the antitumour action of dimethyltriazenes. Cancer Chemother Pharmacol 21:241 (1988)

23. E Benfenati, P Farina, T Colombo, G De Bellis, MV Capodiferro and M D'Incalci. Metabolism and pharmacokinetics of p-(3,3-dimethyl-1-triazeno) benzoic acid in M5076 sarcoma-bearing mice. Cancer Chemother Pharmacol 24:354 (1989)

TRIAZENES: THERAPEUTIC CONSIDERATIONS AND PERSPECTIVES

G. Cartei

Divisione di Oncologia, Ospedale Civile di Udine
Piazza Santa Maria della Misericordia, I-33110
Udine, Italy

INTRODUCTION

Since the first reports on the synthesis of triazenes at the end of last century, triazenes received a certain attention by chemists. The demonstration that phenyl- and imidazole-dimethyltriazenes caused antitumor effects in laboratory animals led to a wider examination of the biological and pharmacological properties of this class of compounds. As a result of these earlier studies, DTIC has entered the clinical practice as an antitumor agent active against malignant melanoma, soft tissue sarcomas, Hodgkin's lymphoma, APUD cell tumors, and it has also been used less frequently on other solid tumors.

DTIC is currently used clinically as a purely cytotoxic drug; at the same time, basic research has indicated that this drug has specific interesting activities, for example on the immune system of the host, and on the antigenicity and metastasizing capacity of tumor cells, which might have clinical relevance. Moreover, several problems encountered during the current clinical use of DTIC still require and deserve clarification. As an example, DTIC is given in combination with other drugs which act on the immune system (cytotoxic antitumor drugs or biological response modifiers such as interferons), as DTIC does itself; clinical protocols used are largely empirical, and are not based on the current experimental evidence. Also the optimal dose schedule of DTIC still requires to be accurately defined, as well as drug toxicity, as is the case of liver disfunction in terms of hepatocyte and liver venous system damage.

This contribution will thus illustrate and discuss, in the perspective outlined above, some of the problems currently faced during the use of DTIC in the clinical practice, with reference to the experimental and clinical literature available and to the personal experience of the author.

DTIC DOSE SCHEDULE

In animal experiments, a relative lack of dose schedule

Triazenes, Edited by T. Giraldi *et al.*
Plenum Press, New York, 1990

dependency was observed (1), and the same seems also to apply to man (2). In patients with malignant melanoma which were treated with DTIC every 3-4 weeks with 850 mg/m², a 19 % response rate was obtained which is similar to the 20 % obtained with higher doses ranging up to 1450 mg/m² (2). More recently, in a group of 57 patients with melanoma DTIC at a dose of 850 mg/m² has been confirmed to produce a 18 % partial remission rate (3). However, the experience of Hill et al. should be considered, as they observed that an individual higher hematological toxicity caused by DTIC was associated with a statistically significant longer disease free interval; the comparison was made in patients which did not receive DTIC versus patients treated with the drug without relevant hematological toxicity (COG protocol 7040; adjuvant DTIC therapy after melanoma removal) (4,5). Moreover, the same authors by escalating the DTIC dosage may have produced an increased response rate up to 29 %; the median duration of response or survival remained however unchanged (6). Samson et al. reported that in melanoma patients a low dose of DTIC (650 mg/m² once daily every 3 weeks) showed no reduction in therapeutic effects (7). Since the administration of DTIC in the range 650-850 mg/m² every 3 weeks causes the same results, it might be expected that only a further dose increase up to 1000-1450 mg/² is needed in order to obtain a higher response rate at least in melanoma patients. The higher hematological toxicity could perhaps be balanced by a better outcome at least for adjuvant DTIC chemotherapy, but further studies are needed to ascertain this possibility.

In patients with malignant APUD cell tumors initially treated with DTIC in divided doses of 1250 mg/m² over 5 days every 4 weeks, a subsequent administration of only 650 mg/m² (once daily every 4 weeks) gave the same therapeutic results (8). In patients with metastatic sarcomas, Blum et al. devised a combination chemotherapy including DTIC (400 mg/m² once daily for 2 consecutive days) with no dose escalation (9), and their results give credit to the lack of dose dependency for the effects of DTIC (10). In metastatic carcinoid tumor, Van Hazel et al. (11) used 250 mg/m²/day for 5 consecutive days every 4-5 weeks, and obtained results comparable to those reported by Kessinger et al. using 650 mg/m²/day every 4 weeks (8). Because of the small number of patients included in each of the studies on APUD cell tumors, it is difficult to ascertain the optimal DTIC dose schedule to be employed.

ANTITUMOR ACTIVITY OF DTIC IN COMBINATION WITH OTHER ANTITUMOR DRUGS

The Eastern Cooperative Group compared adriamycin to an adriamycin-DTIC combination in disseminated soft tissue sarcomas. The response rate to adriamycin alone in a group of 99 patients was 18%, as compared to 32 % in a group of 100 patients treated with the drugs in combination (12). Gottleib et al. had previously observed for the same tumor type a 9 % complete response rate, and a 25 % partial response rate, with a median survival of 14 months for responding patients using the combination adriamycin-DTIC. In patients treated with adriamycin alone, the survival was 8 months (13). The therapeutic gain reported for this combination of the two drugs clearly appears simply additive, rather than synergistic; indeed, DTIC alone causes a response rate of 8-12 % against soft tissue sarcomas. The additive effects observed for DTIC and adriamycin are obtained for the same tumor type also using the fairly less toxic adriamycin analog epirubicin (14).

In patients with malignant melanoma, many combinations including DTIC have been tried unsuccessfully (see review [15] on 18 combinations). In particular, DTIC and adriamycin showed no advantage over the 19 % response rate caused by DTIC alone; similarly, cisplatin and cyclophosphamide do not add to the effects of DTIC in malignant melanoma (15). A lack of cisplatin additive effect with DTIC in melanoma has been confirmed by Oratz et al. (16), but recently DTIC with high dose of cisplatin (150 mg/m^2) gave a higher response rate (17).

Although DTIC is largely used in combination with vinblastine, bleomycin and adriamycin (ABVD regimen) in Hodgkin's disease, no definitive demonstration of the utility for the inclusion of DTIC in this multi-drug regimens is available. Among the antitumor drugs which exert an action synergistic with that caused by DTIC, ifosfamide should be quoted, since remarkable remission rates have been reported for patients·with both malignant melanoma or soft tissue sarcoma (18,19).

A possible occurrence of synergism for DTIC with other drugs against different tumors cannot be excluded, but its investigation requires the use of this drug for tumors naturally non responsive to DTIC. A potential candidate for this study could be the small cell lung carcinoma, which is sensitive to adriamycin, cisplatin, cyclophosphamide and VP16, and not fairly responsive to DTIC.

INTRA-ARTERIAL ADMINISTRATION OF DTIC

DTIC (800 mg/m^2) in combination with cisplatin has been administered by intra-arterial loco-regional infusion for the treatment of recurrent malignant melanoma localized in the limbs (20). This treatment appears to lack a rationale because of the likely requirement for hepatic microsomal activation of DTIC. Moreover, the proponent authors did not examine the effects of DTIC alone.

DTIC IN MELANOMA METASTATIC TO CNS

DTIC enters CNS to a low degree, and according to Loo et al. the concentration of DTIC in the brain tissue of the healthy dog during constant infusion reaches the 14 % of the plasmatic level (21).

In spite of this finding, we have observed two episodes of sudden tumor necrosis in CNS melanoma metastasis during DTIC therapy. The first patient was a male caucasian subject aged 55, who had to be operated as an emergency because of an intractable cerebral edema, after receiving DTIC 400 mg/m^2/day for 3 days; the successful neurosurgical operation revealed the presence of a single CNS metastasis which became largely necrotic (unpublished observation).

A second patient was treated with BELD regimen (bleomy-

cin, eldisine, lomustine and DTIC). This 53 years old caucasian male had vomited almost all of the orally administered
lomustine, whereas he received all of the planned i.v. DTIC
(200 mg/m² /day for 5 days). The patient died suddenly, and
postmortem examination revealed necrosis of numerous cerebral metastasis. The necrosis of brain metastasis should be
attributed to DTIC, since the drug capable to cross the
blood-brain barrier (lomustine) was only very partially
received by the patient, and bleomycin and vindesine presumably had a limited role (22).

Although reports based on few cases have to be considered with much caution, our experience indicates the need
for safeguard in administering DTIC to melanoma patients
with brain metastasis.

ANTIEMETIC THERAPY DURING DTIC ADMINISTRATION

The gastrointestinal toxicity of DTIC, i.e. nausea and
vomiting, is almost invariably present and it is more severe
during the first few days of therapy. In our practice (23),
we have developed two simple antiemetic regimens, both
effective during the relatively lower chronic dose schedule
(65 to 100 mg/m² /day on days 1-14) as well as for the intermediate or high intermittent dose (400 mg/m² /day 1-3) of
DTIC. The first one is based on the administration of alizapride orally (50 mg) and intravenously (100 mg over 20 min
100 ml saline infusion) immediately before DTIC infusion; an
optional further oral and/or i.m. dose (50 mg) is given at 4
p.m.. For the high intermittent dose of DTIC alone, or
during the CyVADIC combination, Synachten depot (one vial)
is given i.m. on the first day only, which is followed by
alizapride 100 mg per os plus 100 mg i.v. immediately before
chemotherapy. Further alizapride 100-200 mg i.v. is given
afterward according to the body surface area; this drug is
administered (50 mg per os plus 50 mg i.m.) at 4 and 8 p.m..
With this safe regimen, nausea and vomiting are suppressed
in about 75 % of the patients. No extrapyramidal side-effects have been observed during this treatment, whereas mild
hypotension and diarrhea have been encountered. The latter
is probably mainly related to the total i.v. hydration
usually given to the patients treated with cyclophosphamide.

The excess of formaldehyde produced during the hepatic
biotransformation of DTIC could be related to the gastrointestinal toxicity (nausea and vomiting) and headache which
occurred in heavy drinkers during DTIC therapy under our
observation. The abolition of intake of alcoholic beverages
represents the first effective anti-emetic procedure to be
adopted.

LIVER TOXICITY OF DTIC

Liver toxicity has been reported in very few articles,
in comparison with the large use of DTIC over the world.
Luce et al. (24) reported the occurrence of flu-like syn-
dromes with high temperature and hepatotoxicity during DTIC
treatment. Elevation of liver function tests (bilirubin,
transaminases and alkaline phosphatases) occurred 3-4 days
after drug administration in a small number of patients, and
was rapidly reversible. Ley et al. (25), and Erichsen et al.

(26) have reported liver damage, i.e. veno-occlusive liver disease (Budd Chiari syndrome), during the course of malignant melanoma treated with DTIC, confirming the previous report by Swensson-Beck et al. (27). This hepatic vein thrombosis has to be regarded as a very unusual syndrome, considering the widespread use of the drug in relation to the rarity of reports on this toxic manifestation. The association of DTIC with cisplatin, BCNU and tamoxifen successfully used in melanoma patients has caused a high incidence of deep vein thrombosis with pulmonary embolism (28). However, liver venous thrombosis was not observed in spite of the demonstrated procoagulant toxicity exerted by this drug association.

We have prospectively investigated liver function tests in 14 patients given adjuvant DTIC (400 mg/m^2/day for three days every 21-28 days); no relevant hepatotoxicity has been observed, with the exception of an early mild elevation of some liver tests. It has to be noted that at least 21 days after each DTIC cycle, no toxic effects on the liver were appreciated with an exhaustive liver function examination including the 22 test profile of SMA Technicon (G. Cartei et al., manuscript in preparation). In two cases of early death after a drug combination including DTIC (BELD regimen; bleomycin, vindesine, BCNU and DTIC), the post mortem examination failed to reveal damage to the liver and to the liver venous system (22).

Paulusma De Waal et al. in a case of glucagonoma firstly used streptozotocin; because of cholestatic hepatitis attributed to this drug, treatment with DTIC (250 mg/m^2/day for 5 days, once a month) was initiated; no liver toxicity was observed, and the patients survived at least two years after diagnosis (29).

Intra-hepatic arterial DTIC together with localized hyper-thermia has been given in patients with liver metastasis. No toxicity occurred even in the presence of the hyper-thermic liver conditions (30). Franchi et al. used high doses of DTIC in leukemic patients (20 mg/Kg body weight daily up to a total dose of 60 mg/Kg), and did not report any relevant liver toxicity (31). These authors kept the DTIC solution iced before use, and protected the bottle and drips from daylight during the infusion (E. Bonmassar, personal communication), exactly as we are used to do in our Division. It may be that such a procedure adds to the safe administration of DTIC, by preventing or reducing the DTIC spontaneous decomposition.

During a prospective randomized trial comparing DTIC versus DTIC plus tamoxifen on 116 patients with malignant melanoma, two cases of fatal veno-occlusive thrombosis of the liver have occurred in the DTIC alone group (32).

The frequency of liver toxicity by DTIC remains unsettled: a review of the more recent literature (14,16,18,19,22,28,29,31-38) indicates that out of 394 patients treated with DTIC, two died with Budd Chiari syndrome and 24 had a mild transient elevation of liver enzymes (G. Cartei et al., manuscript in preparation).

The study by Sava et al. (39) in mice treated with DTIC, DM-COOK or cyclophosphamide revealed at histological examination of the livers marked hepatotoxicity for cyclophosphamide, which caused necrosis of the hepatocytes with loss of lobular structures, appearance of pyknotic nuclei and cytoplasmatic vacuolization. The hepatotoxicity of DTIC was also evident, though less pronounced than that of cyclo-phosphamide; liver vascular lesions were not observed.

Blood coagulation was also investigated by Giraldi et al. (40) in mice during DTIC therapy (60mg/Kg); thrombocyte number, platelet aggregation by ADP, prothrombin and partial thromboplastin times, procoagulant activity of tumor cells, and plasma fibrinogen levels were determined. Prothrombin and partial thromboplastin times were prolonged; platelet aggregation was reduced, and platelet number and plasma fibrinogen were unaffected. This observation indicates that DTIC causes only slight variation, which are less pronounced than those expected to induce a Budd Chiari syndrome, although the clinical relevance of this syndrome is not questionable, and its cause is probably unrelated to coagulative disorders.

It has to be noted that the experimental liver toxicity of DM-COOK is markedly less pronounced than that of DTIC (39), and that the hematological and general toxicity of DM-COOK is marginal (41). These findings suggest that prospective clinical studies on both liver toxicity and blood coagulation in relation to antitumor activity should be undertaken, particularly for new triazenes.

DTIC AND IMMUNITY

DTIC induces at maximum tolerated doses immuno-depression in laboratory animals (see other chapters in this volume). On cell mediated immunity, as observed on allogeneic skin graft rejection in mice, the effects of DM-COOK are less pronounced, or negligible, when compared with those of DTIC (I. Hrsak, personal communication). In normal mice, the oral administration of doses up to 800 mg/Kg/day over 21 days caused no appreciable variations in total and differential leukocyte counts; a similar observation was made in mice bearing Lewis lung carcinoma (T. Giraldi, personal communication). On the contrary, DTIC (60 mg/Kg/ day) administered i.p. daily for 11 consecutive days induced a 75 % reduction in total leukocyte (and hence lymphocyte) counts (41).

. Lymphocyte cytotoxicity against autologous or allogeneic melanoma target cells was investigated during DTIC therapy (150-250 mg/m² /day, on days 1 to 5 every 21 days for 3 cycles), and was not suppressed during the first two cycles; during the third cycle, a decline of cytotoxicity was found in 3 out of 10 patients only (42). Also a different dose schedule of the drug (1 mg/Kg b.w./day on days 1 to 10) did not alter the anti-melanoma cytotoxicity in the majority of the treated patients (43).

The effects of DTIC on some immunological tests during a 14 days daily administration of 60 to 160 mg/m² DTIC was recently investigated in man in comparison with a high intermittent dose schedule (250-400 mg/m² over 3-5 days every 3 weeks). With the latter schedule, monocytes, basophils and lymphocytes showed a progressive decrease; after two cycles, a mild sparing effect on OKT₃, OKT₄, OKT₈ and OKT₄/OKT₈ ratio was observed; with the chronic schedule, only a minimal effect on the same tests was appreciated (44,45). On the basis of these results, it appears that such schedule might safely include other drugs non-cross reactive with DTIC, as well as immuno-stimulant compounds.

Various immunological treatments have been added to DTIC in the clinical therapy of malignant melanoma, including the post-operative adjuvant use. These treatments included BCG, *Corinebacterium parvum*, interferons and active immunotherapy with tumor vaccines (46,47). The rationale for the association of DTIC with α2-R-interferon was based on the assumption that DTIC acts as a purely cytotoxic agent with a prevalent effect on G_2 phase of cell cycle (48), whereas α2Rinterferon acts on G_1 (49). However, DTIC causes a complex cytotoxic action on tumor cells (50), and it induces also various immunological modifications on the host as well as on tumor cells.

Kerr et al. used in 19 patients with malignant melanoma DTIC (800 mg/m² once every 4 weeks) in combination with subcutaneous α2aR interferon (α2aRI) $3x10^6$ U/day for 14 days, escalating by $3x10^6$ U/day each week if tolerated; only 1 patient experienced a partial response. No reference was made on previous chemo- and/or immuno-therapy (36). Bajetta et al. used the same dose of DTIC given every 3 weeks, and α2aRI $3x10^6$ U given i.m. daily which were escalated to $9x10^6$ U/day over a period of 6 days for the first 10 weeks and thereafter three times weekly. On 75 previously untreated patients with malignant melanoma and no CNS involvement, 19 (25 %) experienced an objective response (35), which is not very different from that obtainable with DTIC alone. Thomson et al. treated 51 patients with malignant melanoma (5 pre-treated with chemotherapy, 6 with radiotherapy and 7 with adjuvant immunotherapy) with DTIC 200 mg/m² escalating to 400 and 800 every 3 weeks, and with s.c. α2aRI $3x10^6$ U daily on days 1-3 followed by $9x10^6$ U daily on days 4-70 and thereafter twice a week. The results obtained were CR 14 % and CP+PR 32 %; the authors conclude that α2aRI with DTIC has possibly more efficacy than each single agent (34). Vorobiof et al. used in a smaller number of randomized patients with malignant melanoma either DTIC 250 mg/m²/day for 5 days every 4 weeks, or DTIC plus α2bRI $15x10^6$ U/m² s.c. 3 times a week (19 and 18 patients respectively). These authors obtained 7 responses (39 %) in the combination arm versus 3 (16 %) in the DTIC alone arm (42). These four reports differ for the patients' number, for the dose schedule of DTIC and for the type (α2a-, α2b-RI) and dosage of interferon. These results are encouraging, but it is actually impossible to draw a definite conclusion on the quantitative therapeutic benefit that interferons might add to the classic DTIC therapy.

The combination of DTIC with interleukin-2 (IL2) is also currently investigated in numerous cancer Institutions. As a recent example, DTIC (200 mg/m²/24 hr via central vein infusion for 120 consecutive hours every 4 weeks) was given together with IL2 (starting dose of $2x10^6$ U i.v. over 30 min daily on days 1-5 and 8-12, with escalation to $4x10^6$ U if tolerated). In 14 patients with malignant melanoma, 4 PR and 1 CR (35.7 %) were obtained, indicating that the combination is active (37); a larger group of patients should be treated and evaluated before this association, which is toxic and expensive, can be definitely considered active and preferable over other treatment modalities.

From a clinical point of view, it should be recalled that in experimental animals bearing solid malignant tumors and leukemias, DTIC has a limited cytotoxicity which does not totally account for its antitumor action. Although DTIC causes profound effects on host-tumor relationships in experimental animals (see the chapters in this volume by S. D'Atri et al.,P. Puccetti et al. and T. Giraldi et al.), it is not clarified at present if these actions actually participate into the antitumor (and possibly antimetastatic) action of the drug in humans. Indeed, in experimental animals this drug, and its analog DM-COOK, markedly reduce the metastatic behavior of tumor cells and inhibits systemic metastasis by solid tumors and leukemic infiltration of non-hematological organs by leukemic cells. These effects are caused via a non-cytotoxic selective antimetastatic action, and require a direct interaction with the cells in the primary subcutaneous or intramuscular neoplastic focus. When the treatment with triazenes is performed after surgical removal of primary tumor, or on artificial metastasis obtained by i.v. inoculation of tumor cells, it is ineffective. The administration of DTIC after surgical removal of primary tumors, as currently performed for high risk operable malignant melanoma, thus appears questionable in the light of the above reported considerations. This view is strongly supported by the exhaustive review by Lejeune on the clinical ineffectiveness of the adjuvant postoperative use of DTIC for malignant melanoma (47). Since triazenes do not modify the generalized tumor-host relationships in such a way to account for their antimetastatic action when tumor cells have already entered the blood stream, an alteration localized at the level of the primary tumor which inhibits tumor cell intravasation could be operative.

Finally, an interesting observation has been made by Rosenthal et al. (33) on the concomitant administration of DTIC (by continuous infusion) as a radiosensitizer and radiation, and phase II studies on this topic are warranted.

CONCLUSIONS

From the data and considerations presented so far it appears that DTIC, besides its cytotoxic antitumor action, is a drug endowed with other biological and pharmacological actions of potential therapeutic use. Particular manifestations of its biological, and toxic, actions are encountered during its clinical use, and should be carefully taken into consideration. The use of DTIC in combination with biological response modifiers should be better performed on the rational basis of the abundant immunological experimental evidence available for this drug. A similar consideration is applicable to the use of DTIC in combination with cytotoxic antitumor drugs. Secondly, the preoperative use of DTIC in the presence of the primary tumor could be examined with the aim to increase survival by means of its antimetastatic action based on the inhibition of tumor cell intravasation. A further theoretical possibility of clinical investigation for triazenes is their chronic postoperative administration aimed to inhibit the further formation of new metastasis from metastasis. Finally, DTIC analogs, such as the salts of p-(3,3-dimethyl-1-triazeno)benzoic acid, appear so far in laboratory animals to have interesting properties which seem to justify the performance of accurate clinical trials which may lead to the development of advantageous substitutes for DTIC.

ACKNOWLEDGEMENTS The work of the author was supported by Associazione Italiana per la Ricerca sul Cancro (AIRC, Milano), and by Associazione Oncologica Italiana (AOI, Udine).

REFERENCES

1) J.A. Montgomery, Experimental studies at Southern Research Institute with DTIC (NSC45388), Cancer Treat. Rep. 60: 125 (1976).

2) D.H. Cowan, and D.E. Bergsagel, Intermittent treatment of metastatic malignant melanoma with high dose 5-(3,3-dimethyl-1-triazeno)imidazole-4-carboxamide (NSC45388), Cancer Chemoter. Rep. 55: 175 (1971).

3) K.I. Pritchard, I.C. Quirt, D.H. Cowan, D. Osoba and, G.J. Kutas, DTIC therapy in metastatic malignant melanoma: a simplified dose schedule, Cancer Treat. Rep. 64: 1123 (1980).

4) G.J. Hill, G.E. Metter, S.E. Moss, and F.M. Golomb, DTIC therapy for melanoma: correlation of toxicity with response and longevity in 742 patients, AACR Proc. 17: 244 (1976).

5) G.J. Hill, S.E. Moss, F.M. Golomb, and G.E. Metter, DTIC and combination therapy for melanoma, Cancer 47: 25 (1981).

6) G.J. Hill, G.E. Metter, E.T. Krementz, W.S. Fletcher, F.M. Colomb, G. Ramirez, T.B. Grage, and S.E. Moss, DTIC and combination therapy for melanoma. II. Escalating schedule of DTIC with BCNU, CCNU and vincristine, Cancer Treat. Rep. 63: 19 (1979).

7) M.K. Samson, L.H. Baker, G. Cummings, R.W. Talley, B. McDonald, and D.B. Bhathena, Clinical trial of chlorozotocin, DTIC and dactinomycin in metastatic malignant melanoma, Cancer Treat. Rep. 66: 371 (1982).

8) A. Kessinger, J.F. Foley, and, H.M. Lemon, Therapy of malignant APUD cell tumors. Effectiveness of DTIC, Cancer 51: 790 (1989).

9) W. Blum, J.S. Greenberger, G.P. Canellos, and E. Frei, Successful treatment of metastatic sarcomas with cyclophosphamide, adriamycin and DTIC (CAD), Cancer 46: 1722 (1980).

10) L. Nathanson, J. Walter, and J. Horton, Characteristics of prognosis and response to an imidazole carboxamide in malignant melanoma, Clin. Pharmacol. Ther. 12: 955 (1979).

11) G.A. Van Hazel, J. Rubin, and, C.G. Moertel, Treatment of metastatic carcinoid tumor with dactinomycin or dacarbazine, Cancer Chemother. Rep. 67: 583 (1983).

12) E.C. Borden, D. Amato, H.T. Enterline, H. Lerner, and P.P. Carbone, Randomized comparison of adriamycin regimens for treatment of metastatic soft tissue sarcomas, ASCO Proc. 2: 231 (1983).

13) J.A. Gottlieb, L.H. Baker, J.M. Quagliana, J.K. Luce, J.P. Whitecar, J.C. Sinkovics, S.E. Rivkin, R. Brownlee, and E. Frei, Chemotherapy of sarcomas with a

combination of adriamycin and DTIC, <u>Cancer</u> 30: 1632 (1972).

14) M. Lopez, S. Carpano, L. Di Lauro, P. Vici, and E.M.S. Conti, Treatment of advanced soft tissue sarcomas with epirubicin and DTIC, <u>5th Europ. Conf. Clin. Oncol. ECCO5</u>, London 3-5 sept. 1989, Abst. P-0462.

15) M.J. Mastrangelo, A.R. Baker, and H.R. Katz, Cutaneous melanoma, <u>in</u>: "Cancer principles and practice of oncology," V.T. De Vita, S. Hellmann and S.A. Rosenberg, eds., II Edition, Lippincott, Philadelphia, 1371 (1985).

16) R. Oratz, J.L. Speyer, M.D. Green, R. Blum, J. Wernz, D. Roses, M. Harris, and F.M. Muggia, DTIC and cisplatinum chemotherapy in metastatic malignant melanoma, <u>ASCO Proc.</u> 6: 208 (1987).

17) C. Portlock, J. Murren, A. Buzaid, C. Davis, and W. DeRosa, High dose cisplatin and dacarbazine in metastatic melanoma, <u>ASCO Proc.</u> 8: 284 (1989)

18) J. Casal, A.J. Movano, C. Crespo, A. Marques, L. Carbanas, and R. Moreno, Ifosfamide and DTIC in metastatic malignant melanoma. An effective combination, <u>5th Europ. Conf. Clin. Oncol. ECCO5</u>, London 3-7 sept. 1989, Abst. P-0602

19) W. Campbell, O. Beloqui, P. Herranz, M. Santos, M. Sureda, V. Hidalgo, A. Gil, E. Barrajon, and F.A. Calvo, Ifosfamide, doxorubicin, dacarbazine and amphotericin B in metastatic soft tissue sarcoma: preliminary results of a phase II study, <u>AACR Proc.</u> 29: 222 (1988).

20) C. Charnsangavej, V.P. Chuang, S. Wallace, C.S. Soo, and T. Bowers, Angiographic appearance of recurrent malignant melanoma before and after intra-arterial chemotherapy, <u>Radiology</u> 142: 347 (1982).

21) T.L. Loo, E.A. Strasswender, J.H. Jardine, and E. Frei III, Clinical pharmacological studies on 5-(dimethyl-triazeno)-imidazole-4-carboxamide (NSC 45388), <u>AACR Proc.</u> 8: 42 (1967).

22) G. Cartei, T. Ceschia, P. Marsilio, L. Clocchiatti, G. Fasola, G. Morandini, D. Galletti, and A. Sibau, Effectiveness and toxicity of BELD polychemotherapy in advanced malignant melanoma, <u>Tumori</u> 75: 229 (1989).

23) G. Cartei, A. Bononi, G. Interlandi, F. Cartei, and A. Cantone, Alizapride nell'antiemesi durante monochemioterapia con cis-DD-platino, Atti Sec. Congr. Naz. Soc. Ital. Cure Palliative, 16 dicembre 1989, p. 96.

24) J.K. Luce, W.G. Thurman, B.L. Isaacs, and R.W. Talley, Clinical trial with the antitumor agent 5-(dimethyl-triazeno)imidazole-4-carboxamide (NSC 45388), <u>Cancer Chemother. Rep.</u> 54:119 (1970).

25) F. Ley, M. Winzer, M. Weber, and M. Hypa, Budd-Chiari-Syndrom nach Dacarbacin (DTIC) therapie bei malignem Melanom - ein vermeidbarer Zwischenfall, <u>Z. Hautkr.</u> 60:961 (1985).

26) C. Erichsen and P-E. Jonsson, Venoocclusive disease after dacarbazine therapy (DTIC) for melanoma, <u>J. Surg. Oncol.</u> 27: 268 (1984).

27) H. Swensson-Beck and W.H. Trettel, Budd Chiari syndrom bei DTIC-therapie, <u>Hauzart</u> 33: 30 (1982).

28) E.F. McClay, M.J. Mastrangelo, R.E. Bellet, and D. Berd, An effective chemo hotmonal therapy regimen for the treatment of disseminated malignant melanoma, ASCO Proc., 6:208 821 (1987).

29) J.H. Paulusma-DeWaal, F.T. Bosman, H.R.A. Fischer, P.C. Van der Velden, and W.H.L. Hackeng, The glucagonoma syndrome, Neth. J. Med. 25: 127 (1982).

30) F.K. Storm, L.R. Kaiser, J.E. Goodnight, W. Harrtson, R.S. Helliott, P.H.D. Antoniette, S. Gomes, and D.L. Morton, Thermochemotherapy for melanoma metastases in liver, Cancer 49: 1243 (1982).

31) A. Franchi, S. D'Atri, E. Bonmassar, D. Piccioni, F. Mandelli, F. Malagnino, M. Masi, and G. Papa, High dose dacarbazine in acute leukemia: preliminary immunotoxicological studies, Farmaci e Terapia 5 (suppl. 5): 63, (1988)

32) G. Cocconi, M. Bella, F. Calabresi, M. Tonato, R. Canaletti, C. Boni, F. Buzzi, G. Ceci, E. Corgna, R. Lottici, F. Papadia, M. Sofra, and M. Bacchi, DTIC Vs DTIC plus tamoxifen in metastatic malignant melanoma – A perspective randomized trial of the italian oncology group for clinical research, 2nd Int. Conf. on Melanoma, Venice 16-19 oct. 1989, Abst. pg. 325.

33) C.J. Rosenthal, A. Ohri and M. Rothman, Phase I study of DTIC by continuous infusion and concomitant radiation therapy, ASCO Proc. 6: 43 (1987).

34) D.B. Thomson, R.C. McLeod, and P. Hersey, Phase I/II study of tolerability and efficacy of recombinant interferon with dacarbazine in advanced malignant melanoma, ASCO Proc. 6: 208 (1987).

35) E. Bajetta, E. Negretti, B. Giannotti, L. Brogelli, I. Brunetti, M.R. Sertoli, M.G. Bernengo, M.C. Sofra, G. Maifredi, G. Zumiani, G. Comella, R. Buzzoni, and N. Cascinelli, Phase II study of interferon alpha 2-a and dacarbazine (DTIC) in metastatic melanoma, ASCO Proc. 8: 286 (1989).

36) R. Kerr, P. Pippen, R. Mennel, and S. Jones, Treatment of metastatic malignant melanoma with a combination of interferon alpha 2-a (Roferon) and dacarbazine (DTIC), ASCO Proc. 8: 288 (1989).

37) N.E.J. Papadopoulos, J. Howard, J.L. Murray, C. Plager, S. Legha, J. Reuben, J.U. Gutterman, and R.S. Benjamin, Phase I-II DTIC and interleukin-2 (IL-2) trial for metastatic malignant melanoma, ASCO Proc. 8: 290 (1989).

38) D.A. Vorobiof and G. Falkson, DTIC versus DTIC and recobinant interferon α2b in the treatment of patients with advanced malignant melanoma, 5th Europ. Conf. Clin. Oncol. ECCO5, London 3-7 sept. 1989, Abst. O-0714

39) G. Sava, T. Giraldi, L. Perissin, S. Zorzet, F. Mallardi, and V. Grill, Infiltration of the liver and brain by tumor cells in leukemic mice: prevention by dimethyltriazenes and cyclophosphamide, Tumori 70: 477 (1984).

40) T. Giraldi, G. Sava, E. Mitri, and R. Cherubino, Hemostasis and mechanism of action of selective antimetastatic drugs in mice bearing Lewis lung carcinoma, Eur. J. Cancer Clin. Oncol. 20: 961 (1984).

41) G. Sava, T. Giraldi, L. Lassiani, and C. Nisi, Antimetastatic action and hematological toxicity of p-(3,3-
dimethyl-1-triazeno)benzoic acid potassium salt and 5-
(3,3-dimethyl-1-triazeno)imidazole-4-carboxamide used
as prophylactic adjuvants to surgical tumor removal in
mice bearing B16 melanoma, Cancer Res. 44: 64 (1984).

42) M.S. Mitchell, M.B. Mokyr, and J.M. Davis, Effect of
chemotherapy and immunotherapy on tumor specific immunity in melanoma, J. Clin. Invest. 59: 1017 (1977).

43) J. Berkelhammer, M.J. Mastrangelo, R.E. Bellet, and
R.T. Prehn, Chemoimmunotherapy increases the lymphocyte
reactivity of melanoma patients, J. Cancer Clin. Oncol.
15: 197 (1979).

44) G. Cartei, T. Giraldi and R. Carrella, Effects of
dacarbazine (DTIC) on immunology in the human. I: high
intermittent dose (HID), International Conference on
Triazenes; Chemical, Biological and Clinical Aspects,
Trieste 28-29 nov. 1989, Abst. P2, in press

45) G. Cartei, T. Giraldi and R. Carrella, Effects of
dacarbazine (DTIC) on immunology in the human. II: low
and intermediate chronic dose schedule, International
Conference on Triazenes; Chemical, Biological and
Clinical Aspects, Trieste 28-29 nov. 1989, Abst. P3, in
press

46) U. Veronesi, C. Aubert, E. Bajetta, G. Beretta, G.
Bonadonna, N. Cascinelli, J. De Marsillac, R.L. Ikono-
pisov, B. Kiss, T. Krementz, F. Lejeune, Z. Mechi, G.W.
Milton, A. Morabito, P. Mulder, P. Pawlicki, J. Priar-
io, P. Rumke, R. Sertoli, R. Tomin, N. Trapeznikov, and
R. Wagner, (WHO collaborating centres for evaluation of
methods of diagnosis and treatment of melanoma) controlled study with imidazole carboxamide (DTIC), DTIC +
Bacillus Calmette Guerin (BCG) and DTIC + Corynebacte-
rium parvum in advanced melanoma, Tumori 70: 41 (1984).

47) F.J. Lejeune, Malignant melanoma, in: "Randomized
trials in cancer, a critical review by sites," M.L.
Slevin and M.J. Staquet, eds., Raven Press, New York
(1986).

48) R.L. Comis, DTIC (NSC 45388) in malignant melanoma, a
propsective. Cancer Treat. Rep. 60: 165 (1976).

49) G.M.B. Durie, L. Clouse, T. Braich, M. Grimm, and A.B.
Robertone, Intereferon alpha2b and cyclophosphamide
combination studies: in vitro and phase I-II clinical
results, Seminars Oncology 13: 84 (1986).

50) G. Zupi, A. Corsi, A. Sacchi, L. Lassiani, and T.
Giraldi, Effects of dimethyltriazenes on in vitro Lewis
lung carcinoma lines with different metastatic capacity, Invasion Metastasis 4: 179 (1984).

ANTITUMOUR IMIDAZOTETRAZINONES: PRODRUGS TARGETED TO THE MAJOR GROOVE OF DNA

Malcolm F. G. Stevens

Pharmaceutical Sciences Institute
Department of Pharmaceutical Sciences
Aston University, Birmingham B4 7ET, UK

INTRODUCTION

Triazene chemistry is full of surprises and this is the fourth chapter on the subject by the present author. The first review[1] concentrated on detailing the biologically – relevant chemistry of mono – alkyltriazenes; the second[2] focused on the antitumour drug DTIC (1) and speculated that knowledge of the properties of this compound might be a springboard to improved agents; the third chronicled the back – ground to the discovery of the imidazotetrazinones[3] and summarised details of the spectacular pre – clinical biological activities of mitozolomide (formerly azolastone) (2) and the reasons behind the selection of temozolomide (3) as a second – generation agent.

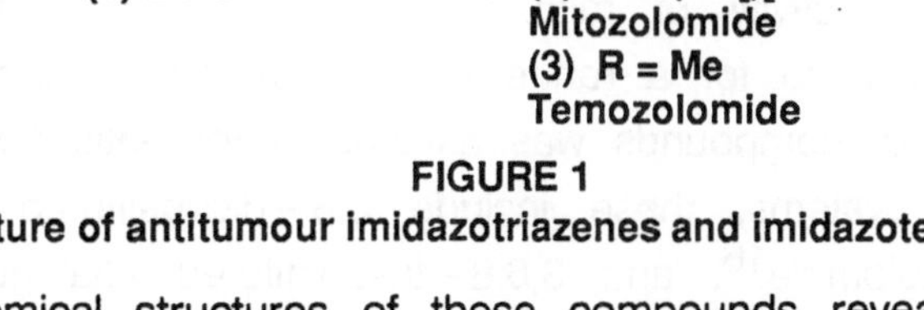

(1) DTIC

(2) R = (CH$_2$)$_2$Cl
Mitozolomide
(3) R = Me
Temozolomide

FIGURE 1
Structure of antitumour imidazotriazenes and imidazotetrazines

The chemical structures of these compounds reveal their triazene lineage (Fig 1) and this chapter will concentrate on summarising the physico – chemical properties of imidazotetrazines in so far as they offer an interpretation of the biological properties of this intriguing group of antitumour drugs.

Triazenes, Edited by T. Giraldi *et al.*
Plenum Press, New York, 1990

It would be wrong to claim that the discovery of antitumour imidazotetrazines owed much to the process of enlightened or rational drug design[3]: rather the most important factor was 'chemical inquisitiveness' within a team with a pedigree in the synthesis of bicyclic compounds with bridgehead nitrogen atoms[4-10] and knowledge of the versatile chemistry of systems with NNN bonds in a cyclic or acyclic environment.[11] Thus it could be predicted that derivatives of the imidazo[5,1-d]-1,2,3,5-tetrazine ring-system (Fig 2) might undergo cleavage at four different bonds - 1,8a; 2,3; 3,4; or 4,5- and that the moieties so liberated might show antitumour activity.

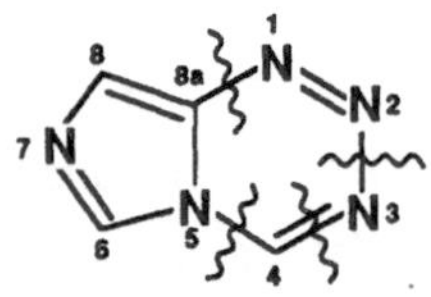

FIGURE 2

Numbering system and sites of bond cleavage in imidazo–[5,1–d]–1,2,3,5–tetrazines

Synthesis of Imidazotetrazinones

Ege and Gilbert[12] devised a synthesis of bicyclic heterocycles comprising an azole ring and a 1,2,3,5-tetrazine ring in 1979. Adaptation of the reaction between diazoimidazoles (4) and isocyanates (5) in work commencing in 1980 at Aston University led to the first synthesis of imidazotetrazinones of general structure (7), <u>via</u> a dipolar intermediate (6) (Fig 3),[13,14] and the identification of the potent antitumour activity of mitozolomide itself.[15] As the synthetic route can be exploited for a range of diazoimidazoles and isocyanates a large portfolio of compounds was assembled for antitumour evaluation in murine test systems; these include 3,8-disubstituted compounds, especially temozolomide[16], and 3,6,8-trisubstituted analogues.[17] The development of a convenient synthesis of the 8-carboxylic acid congener of mitozolomide (8)[18] and its conversion to the corresponding acid chloride (9) opened the door to the preparation of a series of 3-

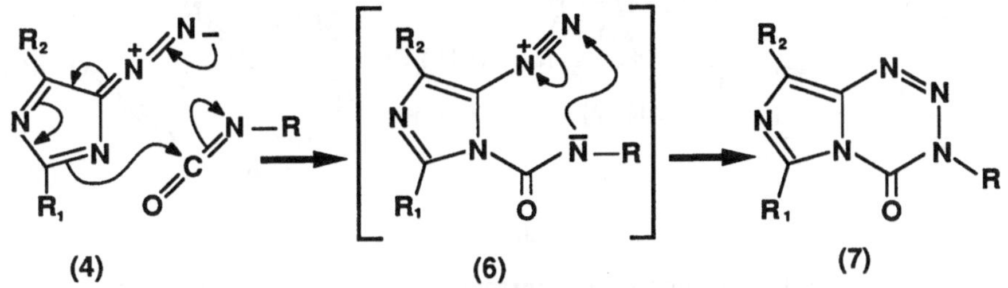

FIGURE 3

Synthesis of imidazo[5,1–d]–1,2,3,5–tetrazinones

chloroethyl – 8 – substituted imidazotetrazinones (10) where the 8 – (X) – substituent is derived from nitrogen, oxygen, sulphur or carbon nucleophiles (Fig 4).

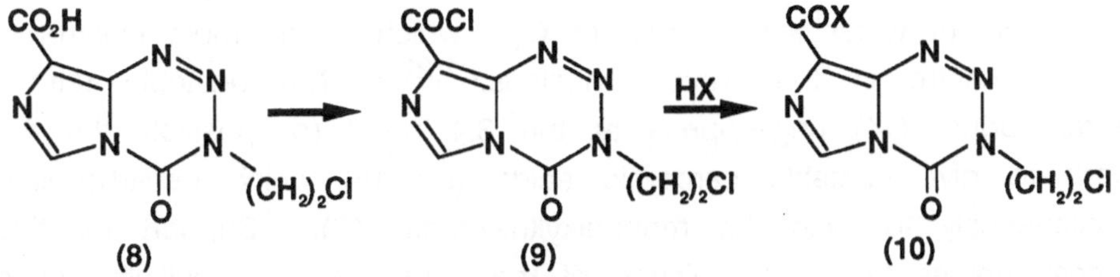

FIGURE 4

Conversion of the carboxylic acid analogue of mitozolomide (8) to derivatives with modified carbonyl substituents

Degradation of Imidazotetrazinones

Although complex, the chemical breakdown of mitozolomide (2) can be subdivided into two general pathways: under thermal and non – aqueous conditions 5 – diazoimidazole – 4 – carboxamide (12) and 2 – chloroethyl isocyanate (13) are regenerated by (formal) cleavage of the 2,3 and 4,5 bonds (Fig 5); in aqueous conditions mitozolomide ring – opens following nucleophilic attack at C(4) to afford the alkylating triazene 5 – [3 – (2 – chloroethyl) – triazen – 1 – yl]imidazole – 4 – carboxamide (MCTIC).[13]

The mechanism of the thermal and non – aqueous pathway possibly involves an initial [1,5] sigmatropic shift to generate an unstable spirobicycle (11) which then undergoes an electrocyclic ring – opening to the diazoimidazole (12) and isocyanate (13). The products isolated from reactions in a range of nucleophiles can be rationalised as arising from interception of these species by the nucleophiles.[13,19]

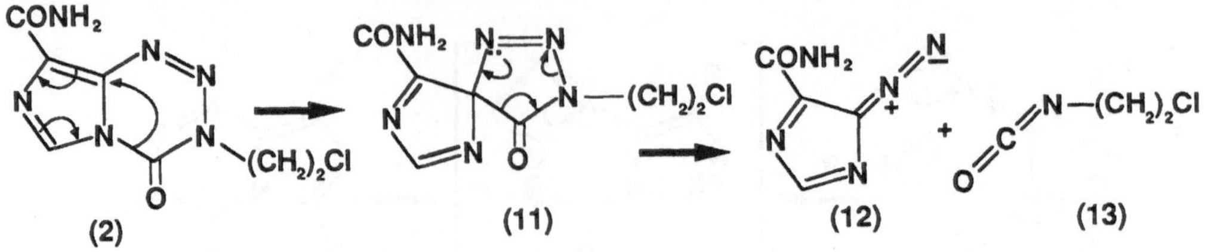

FIGURE 5

Reaction of mitozolomide (2) in thermal and non–aqueous conditions

The decomposition of azolotetrazinones in aqueous conditions is markedly influenced by pH. Unlike 3,3 – dialkyl – 1 – aryltriazenes which are very sensitive to acids,[1] the imidazotetrazinones are stable in acidic but sensitive to basic conditions. At pH 4 mitozolomide has a t½ in buffer of 240 h; at pH 9 the t½ is 0.15 h.[20] A mechanism which accounts for all these observations is summarised (Fig 6). Attack by a molecule of water takes place at C(4) which is the most electron – deficient centre in the bicyclic nucleus.[21,22] The unstable tetra – hedral adduct (14) ring – opens at the 3,4 bond (or possibly the 4,5 bond) to give unstable carbamic acids (eg 15) which decarboxylate spontaneously to yield the monoalkyltriazenes (16). Support for this process comes from the isolation of the appropriate monoalkyltriazenes (16; R = $(CH_2)_2$Cl or Me) when mitozolomide (2)[13] or temozolomide (3)[16]

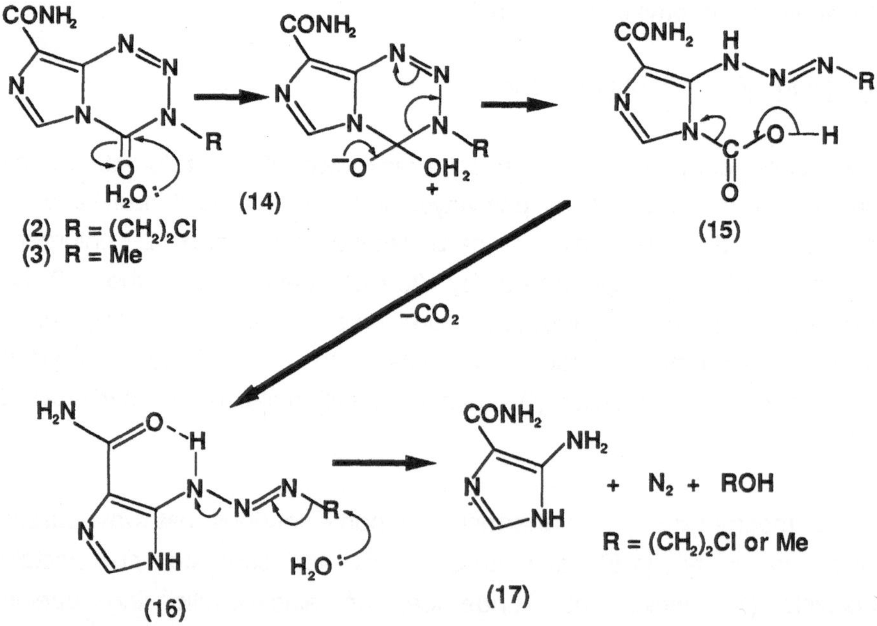

FIGURE 6

Ring–opening of mitozolomide (2) and temozolomide (3) in water

are decomposed in aqueous sodium carbonate solution. Intramolecular H – bonding may freeze the monoalkyltriazenes in the alkylazo tautomeric forms thereby enhancing their potent alkylating properties. Reaction is completed by alkylation of water to liberate 5 – aminoimidazole – 4 – carboxamide (17), nitrogen and an alcohol.[13] The consequences of alkylation of bionucleophiles (eg nucleic acids) is discussed else – where in this book.

STRUCTURE – ANTITUMOUR ACTIVITY RELATIONSHIPS IN IMIDAZOTETRAZINONES

Detailed studies on the antitumour properties of imidazo – tetrazinones have been published in a series of papers[13-18] and are summarised in Fig 7(a). At the 3 – position group R can only be chloroethyl or methyl for potent activity against the TLX5 lymphoma and L1210 leukaemia,[13-16] with chloroethyl preferred to elicit maximum cytotoxicity. The 3 – ethylimidazotetrazinone (7; R = Et, R_1 = H, R_2 = CONH$_2$) has no antitumour properties[16] and, unlike temozolomide (3), does not induce differentiation in K562 human erythroleukaemia cell lines[23] or promote hypomethylation of 5 – methylcytosine levels in DNA.[24]

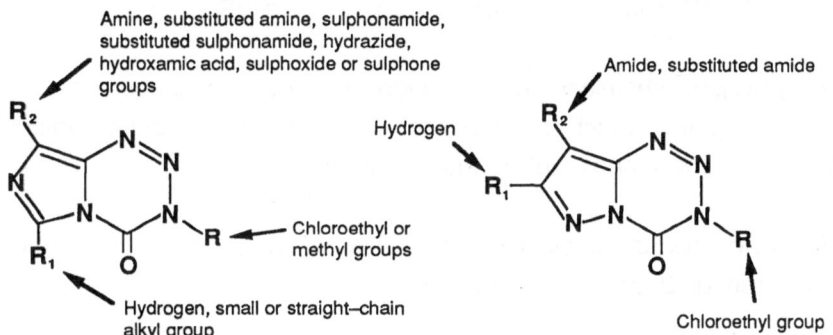

a) Imidazotetrazinones

b) Pyrazolotetrazinones

FIGURE 7

Summary of important structural requirements in (a) antitumour imidazotetrazinones and (b) pyrazolotetrazinones

At C(6) group R_1 should be hydrogen or a small alkyl group to elicit maximum cytotoxicity whereas a larger alkyl group (eg cyclo – hexyl) or aralkyl group (phenethyl) has a dyschemotherapeutic effect.[17] At C(8) (R_2) a rich seam of activity extends through a series of unsubstituted or alkyl or nitro substituted carboxamides, unsubstituted or substituted sulphonamides, hydrazides, hydroxamic acids, sulphoxides and sulphones.[17,18] A recent study of the cyto –

toxicity, stability and metabolism of the 8 – (N,N – dimethylcarbamoyl) analogue of mitozolomide (10; X = NMe$_2$), confirmed that the dimethyl – carbamoyl derivative, although weakly cytotoxic in vitro, is metabolised to the potently cytotoxic monomethylcarbamoyl analogue (10; X = NHMe) in vivo.[25] Nitro, cyano, phenyl, ester, thioester, carboxylic acid or acid azide groups at C(8) abolish or greatly reduce antitumour activity.[17,18] Clearly, ring – opening to a monochloroethyl – triazene alone is not the sole determinant for antitumour activity in this series of compounds – the nature of the C(8) substituent has a critical bearing on biological activity.

A number of pyrazolo[5,1 – d] – 1,2,3,5 – tetrazinones have also been evaluated (Fig 7(b)) and activity appears to parallel that in the imidazotetrazinone field. Thus the 8 – carbamoyl – and 8 – (N,N – dimethyl – carbamoyl) – pyrazolotetrazinones were active in vivo, when a 3 – chloro – ethyl substituent was present, but the 8 – nitro derivative was inactive.[3,17] Cheng and co – workers have confirmed the activity of the pyrazole isomer of mitozolomide and showed, surprisingly, that the corresponding isomer of temozolomide is biologically inert.[26]

STRUCTURAL STUDIES ON IMIDAZOTETRAZINONES

A physico – chemical interpretation of the antitumour activity of imidazotetrazinones must account for the above structure antitumour activity relationships and explain the following facts:

i   Mitozolomide alkylates N(7) residues of guanine bases preferably in a run of 3 or 4 guanines;[27]

ii   The differential cytotoxicity of antitumour imidazotetrazinones toward cell lines proficient (Mer$^+$) of deficient (Mer$^-$) in the capacity to repair guanine O(6) alkylations[28] implies that a chloroethylation or methylation lesion at this site is crucial for antitumour activity;

iii   Mitozolomide and temozolomide are prodrugs which undergo base – promoted conversion to the triazenes (16; R = (CH$_2$)$_2$Cl or Me), respectively (Fig 6). It is unlikely that these latter fugitive moieties could act as pools of circulating and selectively targeted alkylating agents. In this respect these reactive intermediates differ from hydroxymethylmethyltriazenes generated metabolically in the liver from dimethyltriazenes which are (relatively) stable[29] and could diffuse to distant tumour sites.

iv   Mitozolomide and chloroethylnitrosoureas show identical anti – tumour activity, and cross – resistance patterns in rodent tumour test systems.[15]   This preclinical similarity[30] extends to the clinical experience where the dose – limiting side effects of mitozolomide and chloroethylnitrosoureas is a profound and protracted bone marrow suppression, notably thrombo – cytopenia.[31 – 34]

Guanine N(7) and O(6) sites reside in the major groove of DNA and the new proposal that the stable intact bicyclic imidazotetrazinone prodrugs achieve access to the major groove is worthy of detailed evaluation.

The X – ray crystal structure of mitozolomide has been determined by Lowe and co – workers.[35]   The asymmetric unit contains two independent molecules which are 180° rotamers about the C(8) – C(81) bond (Fig 8). In one rotamer (a) the coplanar carboxamide group makes an NH....N hydrogen bond to N(1) of the tetrazine ring; in the other (b) the hydrogen bond is to N(7) in the imidazole ring. In the case of temozolomide the barrier to rotation about the ring to carboxamide bond has been calculated by the <u>ab initio</u> quantum mechanics programme Gaussian 80 and found to be 27.4 kJ/mol.[22]   The rotamers should therefore be fully interconvertible under physiological conditions.

Although mitozolomide and temozolomide are the only two imidazo – tetrazinones to be evaluated clinically, most highly – active analogues have at least one hydrogen bond donor in the C(8) substituent which could contribute to base pair recognition (at cytosine?) in the major groove.[36]   There are exceptions to this rule however. The potent <u>in vivo</u> activity of the dimethylcarboxamide (10; X = NMe$_2$) has been mentioned previously and can be explained by invoking metabolic activation to the monomethylcarboxamide (10; X = NHMe) which does have an hydrogen bond NH donor.[25] Presumably, a similar argument accounts for the <u>in vivo</u> activity of the 8 – (N,N – dimethylsulphamoyl) analogue of mitozolomide,[17] but the correspondingly potent activities of derivatives with sulphone or sulphoxide substituents at C – 8 may require an alternative hydrogen – bonding mode with the oxygen atom acting as acceptor. Inactive compounds such as the 8 – cyano, 8 – nitro and 8 – phenyl imidazotetrazinones with 3 – chloroethyl substituents would be unable to participate in hydrogen bonding interactions. The

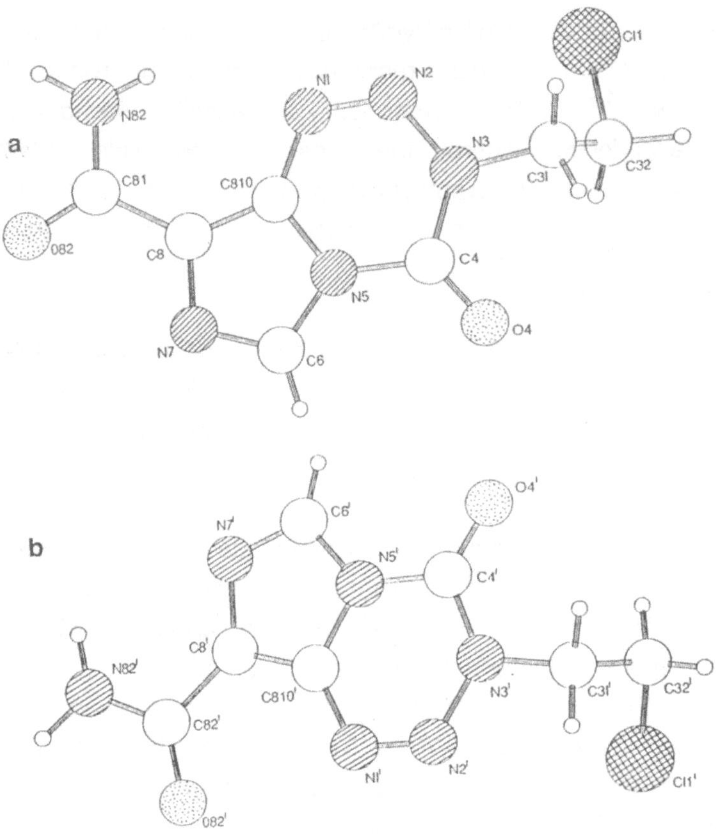

FIGURE 8

Orientations of the carboxamide groups in the crystal structure of mitozolomide. Rotamer (a) has an NH...N hydrogen bond to N (1) of the tetrazine ring; rotamer (b) has a hydrogen bond to N (7) of the imidazole ring. Oxygen atoms are stippled, nitrogen atoms hatched, and chlorine atoms cross – hatched.

presence of a bulky substituent at C(6) may interfere either sterically, with binding to a major groove recognition site, or impede access of a nucleophile to C(4) which initiates ring – opening and presages conversion to the bioactive alkylating species.

A model which encompasses all the above observations relevant to the activity of imidazotetrazinones can be constructed (Fig 9). The novel feature of the model is the proposal that a sequence of guanine residues, together with associated water molecule(s) in the major groove of DNA, provides the nucleophilic micro – environment to activate the intact bicyclic imidazotetrazinone prodrugs.

To satisfy this hypothesis the initial binding of the prodrugs in the major groove must be augmented by hydrogen bonding encounters involving the substituent at C(8), which, in active compounds, is aligned in a coplanar orientation with respect to the bicyclic nucleus.[35,36] Nucleophilic epicentres at N(7) or O(6) in a run of guanines could then 'activate' a molecule of water thereby facilitating orthogonal nucleophilic addition by water at C(4) of the imidazotetrazinones as depicted (Fig 9a). The role of the guanine

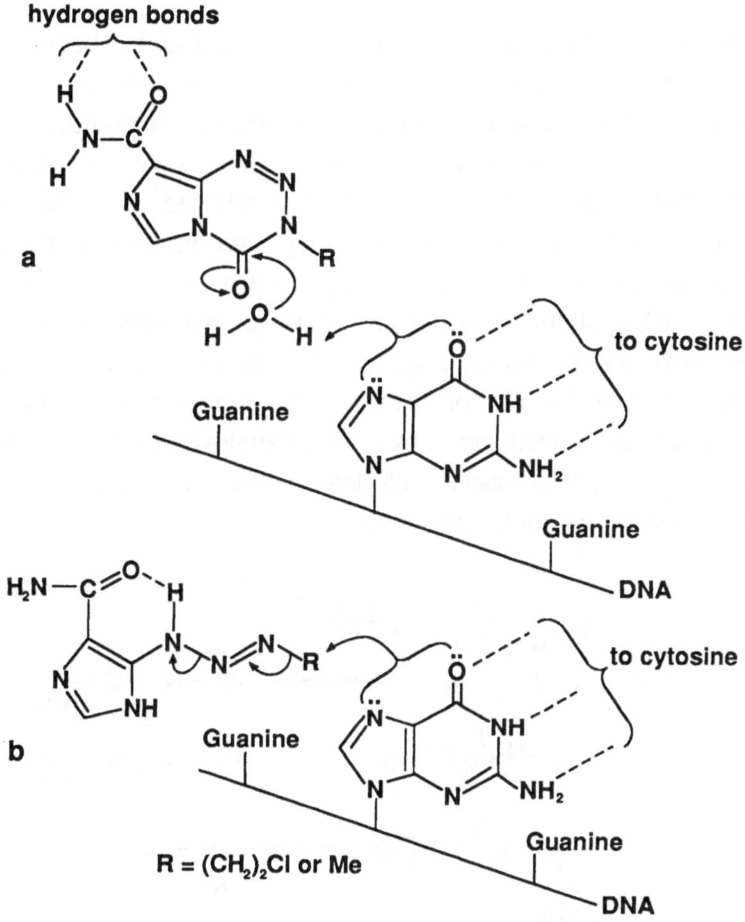

FIGURE 9

Activation of imidazotetrazinones in the major groove of DNA (a) and alkylation of guanine residues (b) by monoalkyltriazenes

residue(s) could be that of neighbouring – group base catalyst similar to that of the imidazole ring of histidine amino acids in serine proteases which are themselves activated by buried aspartate residues.

In the tetrahedral adducts formed following water attack, it is predicted[22] that a significant lengthening and weakening of the N(3) – C(4) and C(4) – N(5) bonds would lead to an energetically favoured ring – opening, loss of a molecule of carbon dioxide and unmasking of the alkylating character of the bioactive monoalkyltriazenes (see also Fig 6). Generation of such reactive entities within the rut of the major groove of DNA would be expected to lead to an S_N2 alkylation of nucleophilic residues in the immediate vicinity, ie N(7) and/or O(6) sites of guanine (Fig 9b).

Nitrosoureas, which are also prodrugs activated by bases, can be accommodated – tentatively – within the same model. Although nitrosoureas (18) show similar structure – antitumour activity relationships to the imidazotetrazinones with only chloroethyl or methyl groups, but not ethyl groups, allowed at the putative alkylating centres (R in Structure 18), they can tolerate a wide range of substituents (X) attached to the non – nitrosoamide nitrogen atom. These latter substituents range from polar to hydrophobic groups, or conjugates with amino acids, sugars, steroids or heterocyclic bases[37] and it requires a major act of faith to accept that they might bind to similar sequences occupied by imidazotetrazinones. However, nitrosoureas do preferentially alkylate runs of guanines in a comparable manner to imidazotetrazinones.[27]

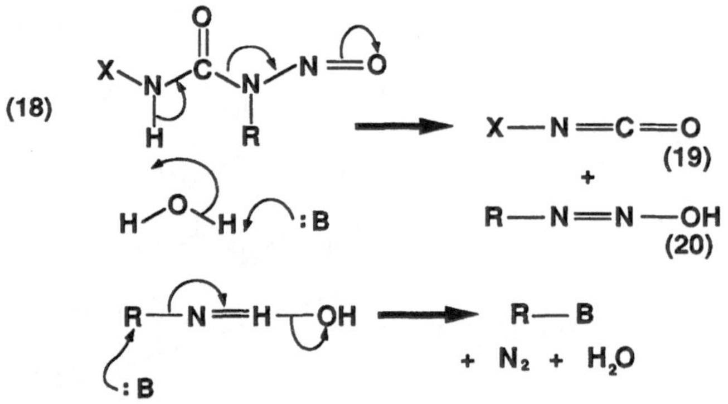

B = Guanine residue in a run of guanines
X = Range of substituents
R = $(CH_2)_2Cl$ or Me

FIGURE 10 Proposed activation of antitumour nitrosoureas and alkylation of guanine residues

To satisfy the model these guanine residues, acting as bases (B in Fig 10), must abstract the N – H atom of the loosely bound nitrosourea (18) through the agency of a water molecule, thereby generating an isocyanate (19) and an alkyldiazohydroxide (20).

Subsequent alkylation of guanine residues at N(7) or 0(6) positions then proceeds by the standard S_N2 reaction. An alternative proposal invoking direct attack of the guanine 0(6) nucleophile at the carbon atom of the imidourea tautomer of nitrosoureas (eg BCNU) without the intervention of major groove – bound water molecules has been proposed recently[38].

2 – Chloroethyl (methylsulphonyl)methanesulphonate (Clomesone, NSC 338947) (21) is subtly different from the imidazotetrazinones and nitrosoureas in its initial reactions with DNA since it is a relatively stable, directly – acting chloroethylating agent, not a prodrug, and therefore alkylates guanine residues at N(7) in a random fashion[28] (Fig 11) as predicted by the new model.

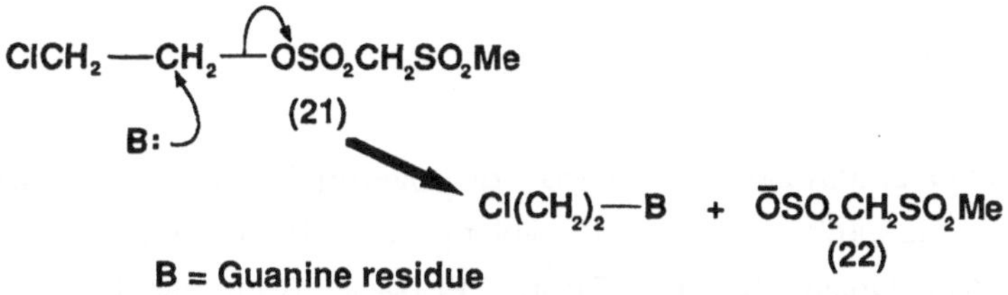

$$ClCH_2-CH_2-OSO_2CH_2SO_2Me$$

(21)

B:

$$Cl(CH_2)_2-B \quad + \quad \bar{O}SO_2CH_2SO_2Me$$

(22)

B = Guanine residue

FIGURE 11

Direct chloroethylation of a guanine residue by clomesone

In the final resort imidazotetrazinone and nitrosourea prodrugs may only differ in their reactions with nucleic acids by the nature of the leaving group attached at the alkylating centres of their activated forms (16) and (20), respectively: these leaving groups are a molecule each of nitrogen and 5 – aminoimidazole – 4 – carboxamide (17) in the case of imidazotetrazinones (Fig 6), and nitrogen and a molecule of water in nitrosoureas (Fig 10). In the case of the directly – alkylating clomesone the leaving group is the (methylsulphonyl) – methanesulphonate anion (22). Subsequent cross – linking of DNA,[28] which can occur with all the chloroethyl – substituted compounds, may then promote clinical sequelae in which marginal clinical activity can be achieved only at the expense of bone marrow suppression with the vital component of favourable therapeutic index missing.[30] The antitumour activity of clomesone in mouse systems[39] shows ominous parallels with activities of mitozolomide and choloroethylnitroso – ureas, and the outcome of the impending Phase 1 trial on clomesone may be equally disappointing. Indeed, following clinical experience with a range of chloroethylating agents which share a common mechanism of action, it is possible to predict both dose – limiting toxicity and maximum tolerated dose of new agents with some confidence, based solely on the molecular weight of the compound. In the case of clomesone the maximum tolerated dose in a single – dose study is likely to be in the range $110-120$ mg m^{-2} with thrombocytopenia as the dose – limiting toxicity.

We are currently designing chemical, biological and computer – modelling studies to test this new hypothesis. It is our hope that the information thus generated will enable us to embark on a rational programme to design new imidaztetrazinones capable of clinically – significant sequence – specific recognition of the topography of more extended stretches of the major groove of DNA.

ACKNOWLEDGEMENTS

The imidazotetrazinones were discovered and developed in a collaborative research programme linking the Pharmaceutical Sciences Institute, Aston University and Rhône Poulenc (formerly May & Baker Ltd) of Dagenham, Essex with additional funding from the Cancer Research Campaign and the Science and Engineering Research Council, UK. The author expresses his special thanks to his colleagues, Dr C E Sansom, Dr C H Schwalbe, Dr K R Horspool and Mr A S Clark whose ideas have been freely drawn upon.

REFERENCES

1.   K. Vaughan and M.F.G. Stevens, Chem. Soc. Review, 7, 377 (1978).

2.   M.F.G. Stevens, DTIC: a springboard to new antitumour agents, in: "Structure – Activity Relationships of Anti – Tumour Agents, D.N. Reinhoudt, T.A. Connors, H.M. Pinedo, and K.W. Van de Poll, eds., Martinus Nijhoff, The Hague, pp. 183 – 218 (1983).

3.   M.F.G. Stevens, Second – generation azolotetrazinones, in: "New Avenues in Developmental Cancer Chemotherapy", Academic Press, London, pp. 335 – 354 (1987).

4. M.F.G. Stevens, J. Chem. Soc., Perkin Trans. I, 1221 (1972).

5. E.J. Gray and M.F.G. Stevens, J. Chem. Soc., Perkin Trans. I, 1492 (1976).

6. E.J. Gray, M.F.G. Stevens, G. Tennant, and R.J.S. Vevers, J. Chem. Soc., Perkin Trans. I 1496 (1976).

7. E.J. Gray, H.N.E. Stevens, and M.F.G. Stevens, J. Chem. Soc., Perkin Trans. I, 885 (1978).

8. G.U. Baig and M.F.G. Stevens, J. Chem. Soc., Perkin Trans I, 1424 (1981).

9. G.U. Baig, M.F.G. Stevens, R. Stone, and E. Lunt, J Chem. Soc., Perkin Trans. I, 1811 (1982).

10. S.P. Langdon, R.J. Simmonds, and M.F.G. Stevens, J Chem, Soc., Perkin Trans. I, 993 (1984).

11. M.F.G. Stevens, Progr. in Medicin. Chem., 13, 205 (1979).

12. G. Ege and K. Gilbert, Tetrahedron Lett., 4253 (1979).

13. M.F.G. Stevens, J.A. Hickman, R. Stone, N.W. Gibson, G.U. Baig, E. Lunt, and C.G. Newton, J Medicin. Chem., 27, 196 (1984).

14 G.U. Baig, M.F.G. Stevens, E. Lunt, C.G. Newton, B.L. Pedgrift, C. Smith, C.G. Straw, R.J.A. Walsh, and P.J. Warren, U.K. Patent Application, 2,125,402A (1984).

15. J.A. Hickman, M.F.G. Stevens, N.W. Gibson, S.P. Langdon, C. Fizames, F. Lavelle, G. Atassi, E. Lunt, and R.M. Tilson, Cancer Res., 45, 3008 (1985).

16. M.F.G. Stevens, J.A. Hickman, S.P. Langdon, D. Chubb, L. Vickers, R Stone, G.U. Baig, C. Goddard, N.W. Gibson, J.A. Slack, C.G. Newton, E. Lunt, C. Fizames, and F. Lavelle, Cancer Res., 47, 5846 (1987).

17. E. Lunt, C.G. Newton, C. Smith, G.P. Stevens, M.F.G. Stevens, C.G. Straw, R.J.A. Walsh, P.J. Warren, C. Fizames, F. Lavelle, S.P. Langdon, and L.M. Vickers, J. Medicin. Chem., 30, 357 (1987).

18. K.R. Horspool, M.F.G. Stevens, C.G. Newton, E. Lunt, R.J.A. Walsh, B.L. Pedgrift, G.U. Baig, F. Lavelle, and C. Fizames, J Medicin. Chem., in press.

19. G.U. Baig and M.F.G. Stevens, J. Chem. Soc., Perkin Trans I, 665 (1987).

20. J.A. Slack and C. Goddard, J Chromatog., 337, 178 (1985).

21. P.R. Lowe, C.H. Schwalbe, C.D. Whiston, and M.F.G. Stevens, J. Pharm. Pharmacol., 37, 136P (1985).

22. A.S. Clark, M.F.G. Stevens, C.E. Sansom, and C.H. Schwalbe, Anti – Cancer Drug Design, in press.

23. M.J. Tisdale, Biochem. Pharmacol., 34, 2077 (1985).

24. M.J. Tisdale, Biochem. Pharmacol., 38, 1097 (1989).

25. K.R. Horspool, C.P. Quarterman, J.A. Slack, A. Gescher, M.F.G. Stevens, and E. Lunt, Cancer Res., 49, 5023 (1989).

26. C.C. Cheng, E.F. Elslager, L.M. Werbel, S.R. Priebe, and W.R. Leopold, J. Medicin. Chem., 29, 1544 (1986).

27. J.A. Hartley, N.W. Gibson, K.W. Kohn, and W.B. Mattes, Cancer Res., 46, 1943 (1986)

28. N.W. Gibson, J.A. Hartley, D. Barnes, and L.C. Erickson, Cancer Res., 46, 4995 (1986).

29. A. Gescher, J.A. Hickman, R.J. Simmonds, M.F.G. Stevens, and K. Vaughan, Tetrahedron Letters, 5041 (1978).

30. J.A. Double and M.C. Bibby, J. Nat. Cancer Inst., 81, 988 (1989).

31. E.S. Newland, G. Blackledge, J.A. Slack, C Goddard, C.J. Brindley, L. Holden, and M.F.G. Stevens, Cancer Treatment Reports, 69, 801 (1985).

32. R.A. Joss, H.J. Ryssel, A.K. Bischoff, A. Goldhirsch, and K.W. Brunner, Cancer Treatment Reports, 70, 797(1986).

33. M. Harding, D. Northcott, J. Smyth, N.S.A. Stuart, J.A. Green, and E.S. Newlands, Br. J. Cancer, 57, 113 (1988).

34. M. Harding, V. Docherty, R. Mackie, A. Dorward, and S. Kaye, Eur. J Cancer Clin. Oncol., 25, 785 (1989).

35. P.R. Lowe, C.H. Schwalbe, and M.F.G. Stevens, J. Chem. Soc, Perkin Trans II, 357 (1985).

36. K.R. Horspool, Ph.D. Thesis, Aston University (1988).

37. R.S. McElhinney, Eur. J. Cancer Clin. Oncol., in press.

38. N. Buckley and T.P. Brent, J. Amer. Chem. Soc., 10, 7520 (1988).

39. D.J. Dykes, W.R. Waud, S.D. Harrison, W.R. Laster, D.P. Griswold, Y.F. Shealy, and J.A. Montgomery, Cancer Res., 49, 1182 (1989)

O^6-ALKYLGUANINE-DNA-ALKYLTRANSFERASE GENE EXPRESSION AND THE CYTOTOXICITY OF TRIAZENES

G.P. Margison, L.C. Harris, L. Cernakova

Carcinogenesis Department
Paterson Institute for Cancer Research
Christie Hospital, Manchester, M20 9BX   UK

V. Vlckova

Department of Genetics
Comenius University, 842,15   Bratislava
Czechoslovakia

J. Brozmanova, K. Kleibl, M. Skorvaga

Cancer Research Institute
Slovak Academy of Sciences   81232   Bratislava
Czechoslovakia

SUMMARY

DNA repair gene transfection has been used to examine the contribution of specific alkylation products in DNA to the toxic effects of triazenes. Expression of the E.coli O^6-alkylguanine (O^6-AlkG)-DNA-alkyltransferases (ATase) encoded by ada and ogt has been achieved in Chinese hamster lung fibroblasts by means of mammalian cell expression vectors.   In comparison with parent vector transfected control cells such cells are more resistant to killing by the chloroethylating triazeno compound mitozolamide indicating that O^6-AlkG and/or O^4-alkylthymine (O^4-AlkT) are critical lesions. These results support the idea that in normal and in tumour cells, endogenous ATase gene expression may be a major factor in resistance to such agents.

Yeast are inherently more resistant to the toxic effects of mitozolamide than mammalian cells however, expression of the alkylphosphotriester (AlkP) ATase encoded by a truncated version of the ada gene increased their resistance to the toxic effects of this agent. This indicates that, at least in the yeast strain used in these experiments, chloroethyl or possibly hydroxyethyl phosphotriesters are toxic DNA lesions.

INTRODUCTION

Alkylating agents are a diverse group of chemicals which include N-nitroso compounds, sulphonates, sulphates, hydrazines and some triazeno compounds. These agents produce a wide range of biological effects in pro- and eukaryotes including mutation, carcinogenesis, teratogenesis and cytotoxicity and this latter effect has been exploited in the development of antitumour agents. The mechanism(s) by which these effects occur has been studied extensively and considerable evidence has accumulated to indicate that interaction with DNA is the critical event. Alkylating agents can react with nitrogen and oxygen atoms in DNA and 13 different products of monofunctional alkylation have been identified. The relative amounts of the products initially formed in DNA varies considerably according to the characteristics of the alkylating species and it has also been found that many of these products are the substrates of DNA repair enzymes that have presumably evolved to reduce the detrimental effects of environmental or enodgenously generated alkylating agents (1-9). A number of in vitro and in vivo experiments have produced direct or correllative evidence that O^6-AlkG is a major mutagenic, toxic and transforming lesion in DNA. The evidence for this is reviewed elsewhere (9).

One of the many approaches to investigating the relative contribution of O^6-AlkG to the biological effects of alkylating agents has been to use cell lines that have been shown to differ in their capacity to repair this lesion (e.g. see ref 10). A major problem in assessing the results of such experiments is that possible contribution(s) of occult differences between the cell lines is very rarely considered. To overcome such problems we have adopted an alternative approach which involves transfer of the cloned E.coli O^6-AlkG repair genes ogt and ada into repair deficient cells so that the resultant cells will differ only in their capacity to repair defined alkyl lesions in DNA.

The present report summarises some of the work with mammalian cells and includes recent experiments in which expression of an E.coli DNA repair gene has been acheived in yeast.

DNA REPAIR GENE TRANSFER INTO MAMMALIAN CELLS

There are a number of reports of the biological effects of expression of the E.coli ada gene, or truncated or mutated versions of it, in a range of mammalian cells. These have recently been reviewed (9). The present report deals mainly with comparative experiments using ogt and a truncated version of ada that encodes only the O^6-AlkG ATase function.

Isolation and characterisation of ogt and ada

The repair of O^6-AlkG in DNA in prokaryotes and eukaryotes is carried out principally by O^6-AlkG DNA ATases which act by transferring the alkyl group from the O^6-position to a cysteine residue at the active site of the enzyme. The autoinactivating stoicheiometric reaction restores the guanine residue to its predamaged state (11-13). A rapid and sensitive assay based on this mechanism was devised and used to screen an E.coli genomic DNA library from which two genes encoding ATases were isolated (14) one of these encoded a 37kDa dual function ATase that transferred alkyl groups from the O^6-position of guanine and the O^4-position of thymine to one active site cysteine residue (15,16) whilst a separate active site transferred alkyl groups from one of the stereoisomers of alkylphosphotriesters (14-17). The gene responsible was found to be the ada gene which controls the adaptive response in E.coli in which exposure to low non-toxic doses of the methylating agent N-methyl-N-nitro-N-nitrosoguanidine (MNNG) increases their resistance to the toxic and mutagenic effects of a subsequent dose of this agent (18). The ada gene had been cloned earlier by mutant rescue but it had not immediately been realised that it encoded a dual ATase (19).

The other gene had not been cloned previously and was named ogt (20). Ogt encodes a 19kDa O^6-AlkG ATase which differs from the ada encoded enzyme in that it acts on higher O^6-alkylated guanine residues in DNA and on O^4-MeT much more rapidly than the ada protein (16). In unadapted E.coli, ogt expression is around 8x higher than ada expression: unlike ada, ogt appears not to be induced by exposure to adaptive doses of MNNG (21). Amino acid homology between ada (22) and ogt and the constitutively expressed B.subtilis O^6-AlkG ATase gene Dat 1 (23) is extensive especially around the active site of the ada O^6-AlkG ATase protein (Figure 1).

```
Ogt     1 MLRLLEEKIATPLGPLWVICDEQFRLRAVEWEEYSERMVQLLDIHYRKEG 50
Dat 1   1 ..MNYYTTAETPLGELIIAEEEDRITRLFLSQE......DWVDWKETVQN 42
Ada   189 ...VRYALADCELGRCLVAESERGICAILLGDD......DATLISELQQM 229


Ogt    51 .YERISAINPGGLSDKLRDYFAG..NLSIIDTLPTATGGTPFQREVWKTL 97
Dat 1  43 .TEHKETPNLAEAKQQLQEYFAG..ERKTF.SLPLSQKGTPFQQKVWQAL 88
Ada   230 FPAADNAPADLMFQQHVREVIASLNQRDTPLTLPLDIRGTAFQQQVWQAL 279


Ogt    98 RTIPCGQVMHYGQLAEQLGRPGAARAVGAANGSNPISIVVPCHRVIGRNG 147
Dat 1  89 ERIPYGESRSYADIAAAVGSPKAVRAVGQANKRNDLPIFVPCHRVIGKNS 138
Ada   280 RTIPCGETVSYQQLANAIGKPKAVRAVASACAANKLAIVTPCHRVVRGDG 329
                                                      ↑
                                       Alkyl group acceptor

Ogt   148 TMTGYAGG.VQRKEWLLRHEGYLLL.. 171
Dat 1 139 ALTGYAGSKTEIKAFLLNIERISYKEK 165
Ada   330 SLSGY.RWGVSRKAQLLRRE.AENEER 354
```

Figure 1   Amino acid sequences of the ATase proteins encoded by the ogt (top line) and Dat 1 (middle line) genes and the O^6-AlkG ATase domain of the ada gene (bottom line). Identical amino acids are boxed. The Alkyl accepting cysteine residue of the ada protein is indicated.

Construction of Mammalian cell expression vectors

In order to obtain expression of the entire ada gene, truncated versions of it that encoded either the O^6-AlkG or the alkylphophotriester ATase coding regions or the ogt gene in mammalian cells, the majority of the 5' and 3' untranslated regions were eliminated. This was usually achieved by means of suitable restriction endonucleases and intermediate cloning vectors (24,25). In the case of ogt, site directed mutagenesis and linkers designed for optimum expression were used in procedures to be described in detail elsewhere (Harris et al in preparation). The mammalian cell expression vector used in the experiments described here was pZipneoSV(x)1(abbreviated to pZip) and the constructs generated are shown in Figure 2.

Expression of the ATase genes occurs by virtue of the promotor in the 5' LTR and the polyadenylation signal in the 3' LTR. In the unspliced message, the ATase gene is translated: the neo gene might also be translated if read through occurs. The spliced message encodes only the aminoglycoside phosphotransferase that confers resistance to the antibiotic G418 (Geneticin). The disadvantages of splicing- dependant expression are outweighed by the packageability of the vector to produce infective virus and the avoidance of cotransfections with selectable marker genes: every G418-resistant clone should contain the gene in question.

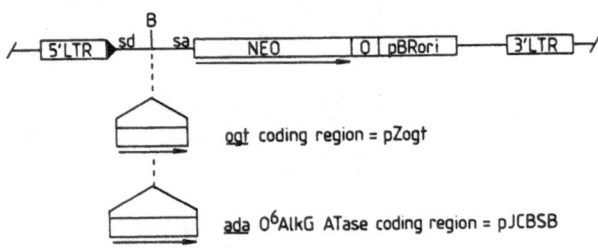

Figure 2 Construction of Moloney murine retrovirus-based mammalian cell expression vectors. The top line shows part of vector pZipneoSV(X)1 which contains the 5' long terminal repeat (LTR) containing the promotor region and the 3' LTR which encodes the polyadenylation signal. Splice donor (sd) and splice acceptor (sa) sites span the unique BamH1(B) cloning site into which ogt or ada ATase coding regions were inserted to generate pZogt or pJCBSB respectively. The vector also contains the aminoglycoside phosphotransferase- encoding neo gene (NEO) an SV40 origin of replication (O) and a pBR233 origin of replication (pBRori). Arrows indicate the direction of translation.

Expression of ogt and ada O^6-AlkG ATase in chinese hamster cells

Vector DNA was introduced into the chinese hamster V79 strain RJKO by transfection using the calcium phosphate procedure in the case of pJCBSB (24,25) or by lipofection in the case of pZogt and pZip. Clones of cells that were resistant to G418 (0.5mg/ml for pJCBSB or 2.0mg/ml for pZogt or pZip) were picked and expanded for further study.

Cell sonicates were prepared and assayed for O^6-AlkG ATase by measuring transfer of [^{3}H]-methyl from methylated substrate DNA to protein as described elsewhere (26). Of 10 pZogt transfected clones, 3 expressed background levels of ATase (2 fm/mg) and in the others activity ranged from 17 to 300 fm/mg. The 3 RJKO clones that were further characterised were 6E (pZip lipofected) SB (pJCBSB transfected 24,25) and LH2 (pZogt lipofected).

PAGE and fluorography of cell extracts showed LH2 cells to contain a single 19kDa ATase which is the size of the ogt encoded protein whilst SB cells contained a major 26 kDa band and several smaller minor bands that are probably the result of alternative in-frame intiation codons (Figure 3). The ATase activity in parent vector transformed cells (6E) was so low that no band has ever been detected on fluorography. However, RJKO-derived cells in which the endogenous ATase gene may have been amplified contain a 24kDa ATase (unpublished results): increased ATase activity in LH2 and SB was therefore not the result of upregulation of the endogenous gene.

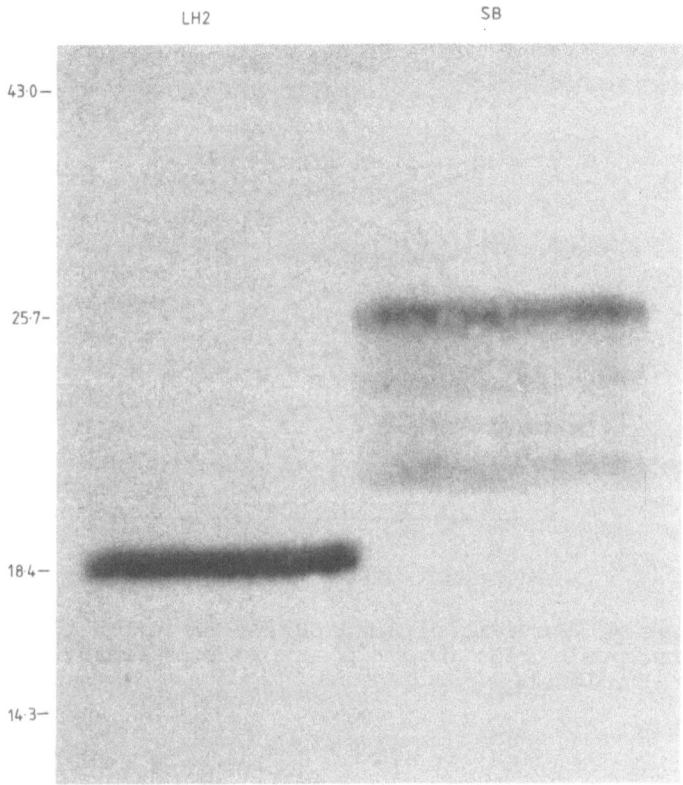

Figure 3   Fluorograph of extracts of LH2 and SB cells following incubation with [³H]-labelled substrate DNA, PAGE and electroblotting to nitrocellulose. The positions of the radiolabelled molecular weight markers are shown on the left.

Analysis of DNA [^{32}P] from the clones by dot-blots or Southern blots using appropriate labelled probes showed LH2 and SB to contain very few (possibly only one) copies of the corresponding ATase gene. Clone 6E contained neo DNA but the copy number has not been determined (Data not shown).

Toxicity of Mitozolomide in LH2, SB and 6E

The survival of LH2, SB and 6E cells after exposure to increasing doses of mitozolomide was determined using a microtitre-plate based (MTT) assay (Figure 4). The culture media were not changed after mitozolamide addition.

As found previously (24,25) SB cells were more resistant to to mitozolamide than control (6E) cells however LH2 cells were found to be considerably more resistant than SB cells. ATase assays of extracts of the cells used in this experiment showed 6E, SB and LH2 to contain 2, 127 and 207 fm/mg ATase. These differences may explain the different degree of protection conferred but it may also indicate that the ogt ATase has a greater capacity to repair potentially lethal lesions than does the ada ATase. (See above and below).

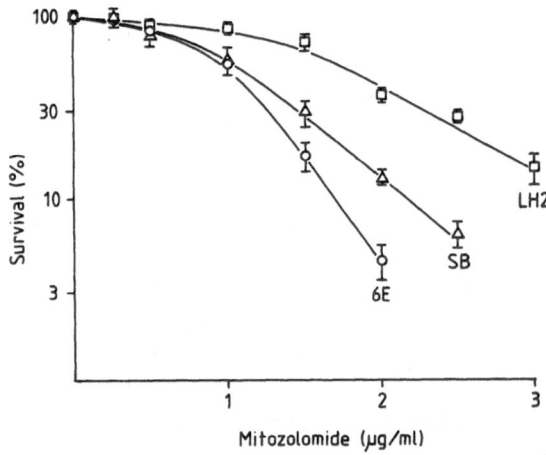

Figure 4 Survival of LH2, SB and 6E (pZipneoSV(x)1 transfected) cells after exposure to increasing doses of mitozolamide.

DNA REPAIR GENE TRANSFER INTO YEAST

Yeast extrachromosomally replicating vectors have recently been used to achieve expression of the protein coding region of ada in endogenous ATase deficient yeast strains (27). Such yeast are more resistant to the toxic and mutagenic effects of MNNG indicating that ada ATase-repairable lesions have similar bilogical effects in yeast and mammalian cells. Here the question of the possible role of chloroethyl phosphotriesters in the toxic effects of mitozolamide in yeast is addressed.

Construction of Yeast expression vector

The protein coding sequence of the ada gene was truncated to an Rsa 1 site upstream of the active site cysteine-encoding region and an in-frame termination codon-containing oligonucleotide ligated at the 3' end. These procedures will be described in detail elsewhere. The resulting construct (PTMT) was ligated into the yeast expression vector pADH to generate pADHPTMT (Figure 5). pADH replicates extrachromosomally by virtue of the 2μ region and the URA gene allows selection in uracil deficient medium. In pADHPTMT the yeast alcohol dehydrogenase promotor drives expression of the ATase.

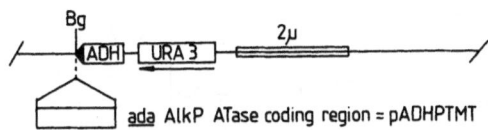

Figure 5 Construction of pADHPTMT, the yeast extrachromosomally replicating vector containing the alkylphosphotriester ATase coding region of the ada gene. The vector was based on YEp24 (see 27) and contains the 2μ replicative element and the URA 3 gene for selection. The promotor was that of the alcohol dehydrogenase gene (ADH) Bg: BglII cloning site.

Expression of the ada Alkylphosphotriester ATase in yeast

Standard procedures were used to transform yeast strain 7799-4B with pADH or PADHPTMT and one clone (N2 and N6 respectively) from each transformation mixture was further characterised. pADH transformed yeast, like their parents contained very low levels of ATase (ca 2 fm/mg protein) wheras pADHPTMT transformed yeast expressed very high levels (ca 1250 fm/mg) and this was shown to be specific for methylphosphotriesters using a substrate that had been depleted of O^6-MeG by prior incubation with excess O^6-AlkG ATase (21).

Indirect evidence for the ability of the truncated ada gene product to act on alkylphosphotriesters in host cell DNA was obtained by treating the yeast with increasing doses of mitozolamide before extraction and assay for ATase activity. Figure 6A shows a dose-dependant decrease in activity although higher doses were relatively much less effective than lower doses.

Toxicity of Mitozolamide in N2 and N6

The survival of N2 and N6 after exposure to increasing doses of mitozolamide was measured by a colony forming assay and the results are shown in Figure 6B. The alkylphosphotriester ATase-expressing clone N6 was considerably more resistant to killing than the pADH transformed clone N2. The shape of these survival curves, characterised by disproportionately lower toxicity at higher doses, is typical of survival curves obtained with MMNG after transformation with the entire protein coding region of the ada gene (27). The reasons for the appearance of the survival curves has yet to be established.

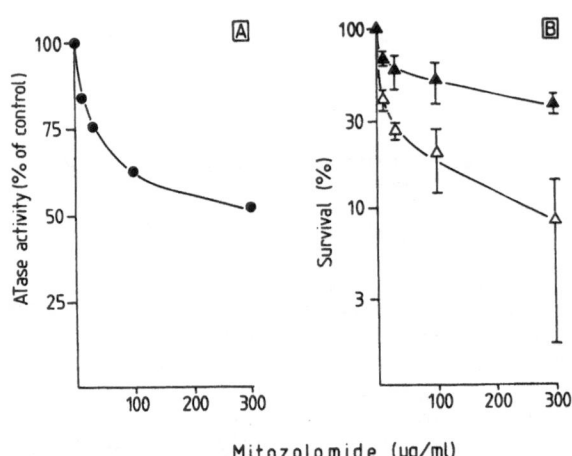

Figure 6   A, Depletion of alkylphosphotriester ATase activity following exposure of PADHPTMT transformed Yeast (clone N6) to increasing doses of mitozolomide. B, Survival of clone N2(,pADH transformed) and clone N6(Δ) yeast after exposure to increasing doses of mitozolomide.

DISCUSSION

The results of the experiments using transfected chinese hamster cells indicates that one or more of the lesions that is repaired by the ogt and ada O^6-AlkG ATases is responsible for a portion of the cytotoxicity produced by the chloroethylating triazene compound mitozolomide. O^6-chloroethylation of guanine in DNA can give rise to interstrand crosslinks in a two-step reaction involving displacement of the halide and cyclisation to $N1-O^6$-ethanoguanine followed by uncoupling at the O^6-position and coupling with the N3 position of cytidine on the opposite strand (28). These crosslinks have been found in DNA after exposure to the related chloroethylating bis-chloroethylnitrosourea (BCNU 29) and they can be prevented by incubation of chloroethylated DNA with cell-free extract containing O^6-AlkG ATase of bacterial (30) or mammalian (31,32) origin. Furthermore mammalian cells that express low levels of endogenous ATase (called mer⁻ or mex⁻ cells) are more susceptible to crosslink formation and the toxicity of chloroethylating agents than cells that express high levels of this enzyme (mer⁻ or mex⁻ cells ref 33). The present results using ogt and ada expressing cells support the idea that toxicity is reduced via repair of O^6-chloroethylguanine and hence prevention of crosslinks. However, it should be considered that similar effects on survival may have been produced by repair of O^6-hydroxyethylguanine or O^4-hydroxy or O^4-chloroethyl thymine if such lesions were cytotoxic. These products would be expected to be repaired by the E.coli ATases albeit much more efficiently by the ogt than the ada encoded proteins. Further studies are in progress in attempts to assess the possible toxic effects of hydroxyethylation of guanine or thymine in comparison with those of chloroethylation.

Another complicating factor in assessing toxic lesions is that mammalian (34) and more recently ada and ogt ATases (T.P. Brent, P. Gonzales, L. Harris and G.P. Margison, unpublished results) bind covalently to BCNU treated oligonucleotides. Although it can be envisaged how such an event would prevent crosslink formation, the precise nature of the covalent complex and the extent to which endogenous repair systems can remove it from DNA and also the consequences of such a repair reaction have yet to be established. It may also be that the exact mechanism of DNA alkylation by BCNU may be pecular to this agent (35) and ATase-DNA covalent linking may ocurr to a much lesser extent after reaction with other chloroethylating agents.

The present work clearly indicates that the endogenous expression of ATase in human tumours may be a critical factor in their response to treatment with alkylating agents. Several groups are now measuring ATase in tumour biopsys (e.g. 36) in order to assess whether or not they will predict tumour response or at least indicate where treatment is likely to be ineffective. Isolation of the human ATase gene should now allow investigation of its tissue specific expression and may provide a mechanism of avoiding some of the toxic side effects of chemotherapeutic alkylating agents.

The preliminary data reported here for yeast indicate that alkylphosphotriesters may also be important toxic lesion in this system. Further work is needed using hydroxyethylating agents in order to assess the relative toxicity of chloroethyl and hydroxyethyl phosphotriesters and similar studies are urgently required in mammalian cells expressing the ada alkylphosphotriester ATase. Although some alkylphosphotriesters have been shown to be repaired slowly in mammalian cells (See 5) there are no reports of an alkylphosphotriester ATase in such cells and the mechanism of this repair remains to be established. Gene transfection may provide a clearer understanding of the importance of alkylphosphotriester toxicity (and repair) and allow the design of more effective antitumour agents.

The present report deals with antitumour aspects of triazeno compounds. In future studies, the mechanisms of carcinogenesis by these agents will be addressed in cell transformation studies and using ada and ogt transgenic animals. Some of the ada (37) and ogt (L Harris unpublished results) transgenic mice we have generated are now showing indications of E.coli ATase expression and such studies should soon be possible.

ACKNOWLEDGEMENTS

Work in the authors laboratory was supported by the Cancer Research Campaign. Thanks to Michelle Boulter for preparation of the manuscript. Mitozolamide was provided by May & Baker Ltd.

REFERENCES

1. Lawley, P.D. (1972) The action of alkylating agents on nucleic acids:
    N-methyl-N-nitroso compounds as methylating agents. In W N Kakahara
    et al (eds) Topics in chemical carcinogenesis, Univ of Tokyo Press,
    Tokyo, pp 237-256.
2. Singer, B. (1975) The chemical effects of nucleic acid alkylation and
    their relation to mutagenesis and carcinogenesis. Prog Nucl Acids Res
    Molec Biol 15, 219-233.
3. Pegg, A.E. (1977) Metabolism of alkylated nucleosides: possible role
    in carcinogenesis by nitroso compounds and alkylating agents. Adv
    Canc Res 25, 195-270.
4. Roberts, J.J. (1978) The repair of DNA modified by cytotoxic,
    mutagenic and carcinogenic chemicals. Adv Radiat Biol 7, 211-436.
5. Margison, G.P. and O'Connor P.J. (1979) Nucleic acid modification by
    N-nitroso compounds. In P.L. Grover (ed) Chemical Carcinogenesis and
    DNA. CRC Press, Boca Raton, pp 111-159.
6. Singer, B. and Kusmierek J.T. (1982) Chemical mutagenesis. Ann Rev
    Biochem 52, 655-693.
7. Lindahl, T. (1982) DNA repair enzymes. Ann Rev Biochem 51, 61-87.
8. Saffhill, R., Margison, G.P. and O'Connor P.J. (1985) Mechanisms of
    carcinogenesis induced by alkylating agents. Biochim Biophys Acta
    823, 111-145.
9. Margison, G.P. and O'Connor P.J. (1989) Biological consequences of
    reactions with DNA: Role of specific lesions. In P.L. Grover and D.H.
    Phillips (Eds) Chemical Carcinogenesis and Mutagensis. Handbook of
    Experimental Pharmacology Vol 94/1. Springer, Heidelberg, In Press.
10. Day, R.S., Babich, M.A., Yarosh, D.B. and Scudiero, D.A. (1987) The
    role of O^6-methylguanine in human cell killing sister chromatid
    exchange induction and mutagenesis: a review. J. Cell Sci. Suppl. 6,
    333-353.
11. Pegg, A.E. (1978) Enzymatic removal of O^6-methylguanine from DNA by
    mammalian cell extracts. Biochem Biophys Res Commun 84, 166-173.
12. Olsson, M. and Lindahl, T. (1980) Repair of alkylated DNA in
    Escherichia coli: methyl group transfer from O^6-methylguanine to a
    protein cysteine residue. J. Biol Chem 255, 10569-10571.
13. Mehta, J.R., Ludlum, D.B., Renard, A. and Verly, W.G. (1981) Repair
    of O^6-ethylguanine in DNA by a chromatin fraction from rat liver:
    transfer of the ethyl group to an acceptor protein. Proc Natl Acad
    Sci (USA) 79, 6766-6770.
14. Margison, G.P., Cooper, D.P. and Brennand, J. (1985) Cloning of the
    E.coli O^6-methylguanine and methylphosphotriester methyltransferase
    gene using a functional DNA repair assay. Nucleic Acids Res 13,
    1939-1952.
15. McCarthy, T.V., Karran, P and Lindahl, T. (1984) Inducible repair of
    O-alkylated DNA pyrimidines in Escherichia coli. EMBO J 3, 545-550.
16. Wilkinson, M.C., Potter, P.M., Cawkwell, L., Georgiadis, P., Patel,
    D., Swann, P.F. and Margison, G.P. (1989) Purification of the E.coli
    ogt gene product to homogeneity and its rate of action on
    O^6-methylguanine, O^6-ethylguanine and O^4-methylthymine in
    dodecadeoxyribonucleotides. Nucleic Acids Res, 17, 8475-8484
17. Weinfeld, M., Drake, A.F., Saunders, J.K. and Paterson, M.C. (1985)
    Stereospecific removal of methylphosphotriesters from DNA by an
    Escherichia coli ada+ extract, Nucleic Acids Res. 13, 7067.
18. Samson, L. and Cairns, J. (1977) A new pathway for DNA repair in
    E.coli. Nature 267, 281-282.
19. Sedgwick, B. (1983) Molecular cloning of a gene which regulates the
    adaptive response to alkylating agents in Escherichia coli. Molec Gen
    Genet 191, 466-472.

20. Potter, P.M., Wilkinson, M.C., Fitton, J., Carr, F.J., Brennand, J., Cooper, D.P. and Margison, G.P. (1987) Characterisation and nucleotide sequence of ogt, the O^6-alkylguanine-DNA-alkyltransferase gene of E.coli.' Nucleic Acids Res. 15, 9177-9193.

21. Potter, P.M., Kleibl, K., Cawkwell, L. and Margison, G.P. (1989b) Expression of the ogt gene in wild type and ada⁻ mutants of E.coli. Nucleic Acids Res. 17, 8047-8060.

22. Morohoshi, R., Hayashi, K. and Munakata, N. (1989) Bacillus subtilis gene coding for constitutive O^6-methylguanine-DNA-alkyltransferase. Nucleic Acids Res. 17, 6531-6543.

23. Demple, B., Sedgwick, B., Robins, P., Totty, N., Waterfield, M.D. and Lindahl, T. (1985) Active site and complete sequence of the suicided methyltransferase that counters alkylation mutagenesis. Proc Natl Acad Sci (USA) 82, 2688-2692.

24. Brennand, J. and Margison, G.P. (1986) Expression of the E.coli O^6-methylguanine-methylphosphotriester methyltransferase gene in mammalian cells. Carcinogenesis 7, 185-188.

25. Margison, G.P. and Brennand, J. (1987) Functional expression of the E.coli alkyltransferase in mammalian cells. J. Cell Sci. Suppl. 6, 83-96.

26. Morten, J.E.N. and Margison, G.P. (1988) Increased O^6-alkylguanine-DNA-alkyltransferase in chinese hamster V79 cells following selection with chloroethylating agents Carcinogenesis 9, 45-49.

27. Brozmanova, J., Kleibl, K., Vlckova, V., Skorvaga, M. and Margison, G.P. (1990) Expression of the E.coli ada gene in yeast protects against the toxic and mutagenic effects of MNNG. Nucleic Acids Res. in press

28. Kohn, K.W. (1977) Interstrand cross-linking of DNA by 1,3 bis (2-chloroethyl)-1-nitrosourea and other 1-(2-haloethyl)-1-nitro-soureas. Cancer Res. 37, 1450-1454.

29. Tong, W.P., Kirk, M.C. and Ludlum, D.B (1982) formation of the cross-links 1-[N³-deoxycytidyl], 2-[N¹-deoxyguanosinyl]-ethane in DNA treated with N, N'-bis (2-chloroethyl)-N-nitrosourea. Cancer Res. 42, 3102-3105.

30. Robins, P., Harris, A.L., Goldsmith, L. and Lindahl, T. (1983) Cross-linking of DNA, produced by chloroethylnitrosourea is prevented by O^6-methylguanine-DNA-methyltransferase. Nucleic Acids Res. 11, 7743-7758.

31. Ludlum, D.B., Mehta, J.R. and Tong, W.P. (1986) Prevention of 1-(3-deoxycytidyl), 2-(1-deoxyguanosinyl)ethane cross link formation in DNA by rat liver O^6-alkylguanine-DNA alkyltransferase Cancer Res 46, 3353-3357.

32. Brent, T.P. (1984) Suppression of crosslink formation in chloroethylenitrosourea-treated DNA by an activity in extracts of human leukemic lymphoblasts. Cancer Res. 44, 1887-1892.

33. Gibson, N.W., Hartley, J.A., La France, R.J. and Vaughan, K. (1986) Differential cytotoxicity and DNA-damaging effects produced in human cells of the mer⁻ and mer⁻ phenotypes by a series of 1-aryl-3-alkyl-triazenes. Cancer Res. 46, 4999-5003.

34. Brent, T.P., Smith, D.G. and Remack, J.S. (1987) Evidence that O^6-alkylguanine-DNA alkyltransferase becomes covalently bound to DNA containing 1,3-bis(2-chloroethyl)-1-nitrosourea-induced precursors of interstrand cross-links Biochem. Biophs. Res. Comm. 142, 341-352.

35. Buckley, N. and Brent, T.P. (1988). Structure-activity relations of (2-chloroethyl) nitrosoureas. 2. Kinetic evidence of a novel mechanism for the cytotoxically important DNA cross-linking reactions of (2-chloroethyl)nitrosoureas. J. Amer. Chem. Soc. 110, 7520-7529.

36. Maynard, K., Parsons, P.G., Cerny, T. and Margison, G.P. (1989) Relationships between cell survival, O^6-alkylguanine-DNA-alkyltransferase activity and reactivation of methylated adenovirus 5 and herpes simplex virus type 1 in human melanoma cell lines. Cancer Res. In Press.

37. Searle, P., Tinsley, J.M., O'Connor, P.J. and Margison, G.P. (1988) Generation of transgenic mice containing a DNA repair gene from E.coli. Proc. Am. Assoc. Canc. Res. 29, 111.

N-METHYLMELAMINES, A UNIQUE CLASS OF ANTI-TUMOUR AGENTS?

I.R. Judson

Drug Development Section, The Institute of Cancer Research, Block E, 15 Cotswold Road, Belmont, Sutton, Surrey SM2 5NG, UK.

HEXAMETHYLAMINE

N-methylmelamines have been investigated as potential anti-tumour agents for nearly forty years (Fig. 1). The parent compound, hexamethylmelamine (HMM) was first synthesised in 1951 in the search for synthetic resins and fibres (1). It was shown at the time that HMM and N^2,N^4,N^6-trihydroxymethylmelamine were cytotoxic versus the Walker 256 tumour in the rat (2,3). However, HMM was not tested in phase I trials until 1965 in which anti-tumour activity was observed against ovarian and lung cancer (4). Nausea and vomiting were dose limiting, additional toxicities being neuropathy and myelosuppression as confirmed by a subsequent study (5). Phase II studies were carried out in the early 1970's and demonstrated broad spectrum anti-tumour activity (6,7).

Clinical Anti-tumour Activity

The activity of single agent HMM in advanced ovarian cancer was found to be comparable with that of the nitrogen mustards. For example in a randomised study, response rates in previously untreated patients with stage III or IV ovarian cancer were melphalan: 36%, 5-fluorouracil: 12% and HMM: 39% (8). In addition, HMM was active in patients whose tumours had proved resistant to alkylating agents (9,10). Because of its relatively marrow-sparing toxicity, lack of cross-resistance and intrinsic activity, HMM was included in many active combinations in the treatment of ovarian cancer (11,12). Unfortunately, however, although attempts were made to 'isolate' the contribution of HMM towards such combination treatment (13,14), the value of this drug in the management of advanced ovarian cancer remains in doubt (15).

Bruckner et al (16), claimed that intensive treatment including HMM gives a survival advantage over regimens lacking this agent. However, where HMM has been used as salvage treatment following cisplatin-containing combination therapy, the results have not been impressive (17). Whether this is due to the fact that cisplatin resistant tumours are cross-resistant to HMM or simply that patients whose disease is resistant to good combination therapy are difficult to treat, is still not clear. A direct comparison between a platinum complex and HMM has not yet been performed.

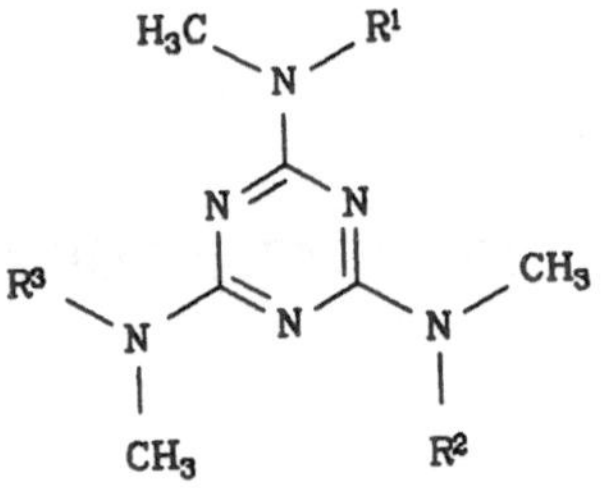

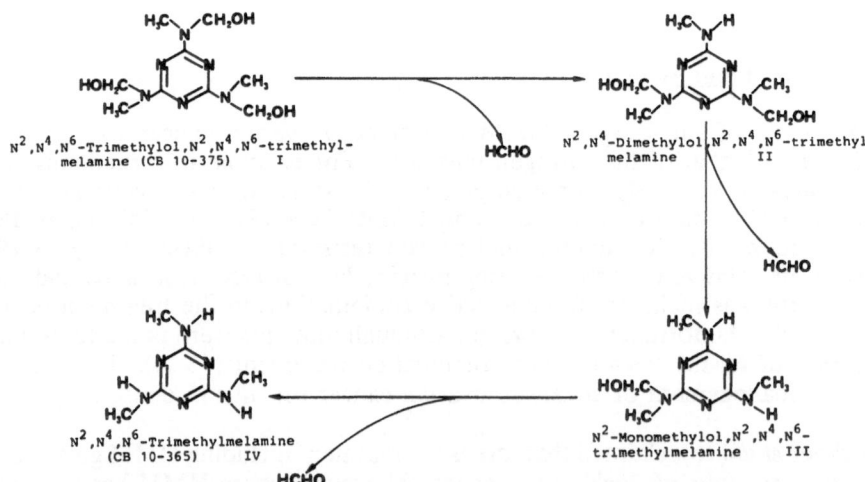

Figure 1
Structure of N-methylmelamines

	R^1	R^2	R^3
Hexamethylmelamine (HMM)	CH_3	CH_3	CH_3
Pentamethylmelamine (PMM)	H	CH_3	CH_3
N^2-monohydroxymethyl-pentamethylmelamine (MHMPMM)	CH_2OH	CH3	CH_3
N^2,N^4,N^6-trihydroxymethyl-N^2,N^4,N^6-trimethylmelamine (Trimelamol)	CH_2OH	CH_2OH	CH_2OH

Figure 2. Route of chemical decomposition of Trimelamol

It was known that methylmelamines could be metabolised by oxidative N-demethylation and that N-hydroxymethylmelamines were unstable, decomposing to release formaldehyde (18).

N-Demethylation was shown to be a major metabolic pathway for melamines in rats and man (19) and the same authors showed that no significant cleavage of the melamine ring occurred *in vivo* (20). Metabolism of HMM may also occur *in vitro* in the presence of liver microsomes (21). Rutty and Connors (22) demonstrated that for a series of N-methylmelamines the degree of demethylation *in vitro* correlated with the *in vivo* antitumour activity of these compounds towards the ADJ/PC6 plasmacytoma. N-hydroxymethylmelamines were also active *in vivo*. In addition, Rutty and Connors used an *in vitro/in vivo* bioassay to demonstrate that N-hydroxymethylmelamines, such as N^2-monohydroxymethyl-pentamethylmelamine (MHMPMM) and N^2,N^4,N^6-trihydroxymethyl-N^2,N^4,N^6-trimethylmelamine (Trimelamol) (Fig.1), were active in the absence of microsomes, whereas HMM was totally inactive under these conditions. Rutty and Abel (23) found that HMM and PMM were cytotoxic towards PC6 plasmacytoma cells *in vitro* but only after prolonged contact, in contrast to the N-hydroxymethyl derivatives which were rapidly cytotoxic.

Although the need for metabolic activation was thus acknowledged, some doubt remained as to the nature of the active species. Rutty and Abel (23) examined the role of the formaldehyde released on decomposition of N-hydroxymethylmelamines, using semicarbazide as a formaldehyde trapping agent. Although semicarbazide pre-treatment virtually abolished the cytotoxicity of MHMPMM and Trimelamol towards L1210 leukaemia cells *in vitro*, PC6 cells were not protected. Since the latter tumour was sensitive *in vivo* this appeared to suggest that the N-hydroxymethylmelamine itself was the active moiety. However, this issue remains unresolved.

WATER-SOLUBLES ANALOGUES OF HMM

The low aqueous solubility of HMM necessitates oral administration. Although an intravenous preparation formulated in a soybean extract has been evaluated (24) this has not yet been shown to have antitumour activity. Oral administration in man results in wide variations in the plasma concentration and half-life of HMM (25). The oral route is also unsatisfactory for a drug which causes marked nausea and vomiting. Therefore a structure-activity investigation of water-soluble analogues of HMM was performed (26). Of the compounds examined, N^2,N^4,N^4,N^6,N^6-pentamethylmelamine (PMM), MHMPMM and Trimelamol (Fig 1) showed significant activity against the PC6 tumour and a human lung tumour xenograft, the P246, grown in immune-deprived mice (27).

Pentamethylmelamine

PMM was subsequently chosen for clinical development. The reasons for this choice included the instability of Trimelamol and MHMPMM, and the fact that PMM was a known metabolite of HMM (20). It was thus presumed that PMM would be active and possess similar toxicity to that of the parent compound in man. It also had a sufficiently high aqueous solubility.

It was therefore a considerable disappointment to find that PMM was extremely toxic, producing dose-limiting nausea and vomiting and profound sedation (28,29,30,31,32). Furthermore, few objective signs of anti-tumour activity were observed. Myelosuppression was generally mild and inconsistent unless the drug was given repeatedly at high doses for 10 days (29). Although dose escalation was usually halted because of gastrointestinal toxicity, one patient died in coma with progressive leukoencephalopathy (28).

Species Differences in Metabolism

The pharmacokinetic studies of Rutty et al (33) demonstrated marked species differences in the metabolic activation of PMM between mouse, rats and humans (mouse > rat >> human). The levels of the putative active metabolites, N-hydroxymethylmelamines, were too low in patients to be detected using the Nash assay, whereas high levels were observed in the mouse. The rat was intermediate in this regard. They therefore proposed that inadequate metabolic activation was the likely explanation for the poor clinical activity of PMM. If this was the case then parenteral administration was possibly detrimental as far as the active metabolites were concerned. By-passing the gut would have abolished first-pass metabolism in the intestinal wall and liver, which may well have resulted in a reduction in metabolic activation. Furthermore, antitumour activity is related to the number of methyl groups (22). In vitro, PMM is less well demethylated than HMM using rat and liver microsomes (34). Therefore, it is possible that PMM is intrinsically less active, although no direct comparison with HMM has been made in man. Morimoto et al (35), found that PMM produced less inhibition of DNA synthesis than HMM in Lieberman plasmacytoma ascites cells in vivo. This tumour was also said to be less sensitive to PMM (36). In general a major difference in activity between HMM and PMM has not been demonstrated in rodents, where N-demethylation is known to be efficient, but failure of metabolic activation could result in such a difference in man.

TRIMELAMOL

Pre-clinical Activity

The case for clinical evaluation of Trimelamol was therefore re-examined. Trimelamol was known to be equally active as PMM against two experimental tumours, namely a human lung tumour xenograft and the murine PC6 plasmacytoma (22). Subsequent work with human ovarian tumour xenografts has demonstrated the activity of Trimelamol against this tumour type (37). Rutty et al (33), had shown that oxidative N-demethylation of PMM was slower in the rat than in the mouse. It was subsequently demonstrated that Trimelamol was superior to PMM versus the Walker 256 sarcoma in the rat (38,39). This provided further evidence for the view that direct administration of a N-hydroxymethylmelamine would be advantageous in man. Finally, toxicological investigations showed that Trimelamol was less neurotoxic than PMM in rodents (40). This was found to be due to the reduced ability of Trimelamol to enter the brain compared with PMM (41).

Formulation

Before phase I trials of Trimelamol could be performed the problem of formulation had to be overcome. The drug is relatively insoluble, chemically unstable and tends to polymerise. A solubility of 9 mg/ml was claimed by Cumber and Ross (26) but the conditions used would have caused considerable decomposition. The maximum aqueous sobility at room temperature of the bulk drug was ≈ 1 mg/ml. A combination of ball-milling to reduce particle size and lyophilisation was used to increase this to 5 mg/ml (39).

Stability was shown to be adversely affected by increased temperature and ionic strength and acid pH (39). Therefore the drug was formulated in 5% dextrose at pH 9.0, all manipulations following the initial dissolution being carried out at 4°C. Polymerisation was shown to be dependent on stability hence this was limited under the conditions used. It has been shown that the major condensation product is, in fact, the dimer (D. Thurston and C. Jackson, personal communication).

The purity of the drug used for the clinical studies was only 92% and it was necessary to show that the major impurity was the primary decomposition product of Trimelamol, N^2,N^4-dihydroxymethyl-N^2,N^4,N^6-trimethylmelamine (39). This would be formed *in vivo* following the administration of Trimelamol and is likely to contribute to the anti-tumour activity of the drug. Therefore, it was not felt that this degree of impurity represented a major problem.

Clinical Studies

A phase I trial was performed at the Royal Marsden Hospital. Two schedules were evaluated, a single IV injection every three weeks and a fractionated schedule of three daily injections repeated every three weeks. Given the known activity of HMM versus ovarian cancer the study was targeted towards this disease. Hence, of a total of 82 patients treated, 58 had advanced ovarian cancer. The dose limiting toxicity proved to be myelosuppression, mainly leukopenia. The maximum tolerated dose (MTD) for the single infusion was 2400 mg/m^2 at which dose the median leukocyte nadir was 1.5 x 10^9/l. Dose related nausea and vomiting were also observed but these were not dose limiting. Above 1100 mg/m^2 gastrointestinal toxicity was usually WHO grade 3 (WHO 1979) on the single dose schedule. A degree of somnolence was observed but no acute sedation, unlike PMM.

The effects of variations in treatment schedule were studied in Balb C mice bearing the PC6 tumour. Dose fractionation resulted in a marked improvement in therapeutic index, as defined by the ratio of 50% lethal dose (LD$_{10}$) / 90% inhibitory dose (ED$_{90}$), rising from 1.8 for a single injection to 4.9 for two daily doses and 3.5 for three daily doses (42). Using the experience from the single infusion schedule it was decided to use three daily doses which allowed a reduction in the individual daily dose below that causing serious emesis. In addition, it was found that a small increase in overall dosage was achieved for the same degree of myelosuppression giving a maximum tolerated dose for the three day schedule of 3000 mg/m^2 with equivalent leukopenia, i.e a median leukocyte nadir of 1.5 x 10^9/l. Some patients crossed over from single dose to three day treatment with a reduction in myelosuppression or on crossover were able to tolerate a dose increase for the same degree of leukopenia. No significant differences in risk factors for myelosuppression nor pharmacokinetic parameters were identified between the two groups of patients which could account for the reduced myelosuppression. This suggests that the improvement was associated with dose fractionation.

Pharmacokinetics

Pharmacokinetic studies were performed at all dose levels and confirmed that unlike PMM, Trimelamol is not subject to species differences in drug clearance. Elimination of the drug was monoexponential and the elimination half life was similar to that of the mouse (39). The median half-life over the dose range 50-3100 mg/m² was 6.5 ± 0.2 min. Plasma clearance was independent of dose with a mean value of 4054.9 ± 1915 ml/min. There was a good correlation between dose and AUC (r = 0.959). There was no significant change in clearance over three days of administration or multiple single infusion and the cumulative area under the plasma concentration/time curve (AUC) was equivalent to that of a single larger dose.

In addition to the relationship between dose fractionation and therapeutic index, there was substantial agreement between the AUC at an effective anti-tumour dose in the mouse, i.e. 90mg/kg, the ED_{90} versus the PC6 tumour, AUC = 2834 µM.min. (1) and the AUC at a toxic or effective anti-tumour dose in man. Myelosuppression was first observed at 1500mg/m² at which dose the mean AUC was 3231 µM.min. Objective anti-tumour responses were seen at the next highest dose of 1800mg/m². The degree of agreement between this rodent tumour model and the behaviour of the drug in man gave weight to the decision to investigate a fractionated dose schedule in the clinic.

Because of the linear relationship between dose and AUC, both in mouse (r = 0.973) and man (r = 0.959), Trimelamol provides a good example of a drug in which to examine the feasibility of pharmacokinetic methods for facilitating dose escalation as suggested by Collins et al (43). They showed that for a number of cytotoxic drugs the AUC at the mouse LD_{10} was a better predictor of the human MTD than the LD_{10} itself; i.e. the ratio: AUC at human MTD/AUC at mouse LD_{10} was closer to unity than the ratio: MTD/LD_{10} (mg/m²). Thus for Trimelamol the AUC at the mouse LD_{10} of 150mg/kg (450mg/m²) was 4355 µM.min. and at the human MTD of 2400mg/ m² was 8543 µM.min giving a ratio of 1.96. On the other hand the corresponding ratio of human MTD/mouse LD_{10} (mg/m²) = 2400/450 = 5.33. Thus for this drug the mouse pharmacokinetic data at a toxic dose provided a much better predictor of the human MTD than did the toxic dose itself. If this information had been used to justify a more rapid dose escalation policy according to the Collins proposals, the dose could have been raised from 50mg/m² to either 400 or 700mg/m² in one step, further escalations being carried out conventionally. Therefore the number of escalation steps might have been reduced from 13 to only 5 or 6, depending on the precise method used. Clearly the saving in terms of time, drug and, more importantly, patients treated at inadequate doseage could be substantial if similar results are obtained in future prospective studies (45).

Clinical Activity in Ovarian Cancer

One complete and 8 partial responses were observed in patients with refractory ovarian cancer. The complete remission was radiological not pathological. In addition there was a durable partial remission in a patient with heavily pretreated Hodgkin's disease. Four of the responding patients with ovarian cancer had progressive disease while on treatment with cisplatin or carboplatin. All objective responsess occurred in patients treated at a dosage of 1800 mg/m² or above. Both schedules appeared equally effective with five responses observed on each.

In a subsequent phase II study in ovarian cancer using the three day schedule at a daily dose of 800 mg/m² a response rate of only 9.5% was observed which was rather lower than expected (45). However all patients included in this study were heavily pre-treated, hence the potential value of this drug in the treatment of ovarian cancer remains uncertain. Further phase II evaluation in this and other tumour types has been hampered by the fact that attempts to scale up the formulation method used for the studies described here have so far proved unsuccessful.

Does Metabolic Degradation Occur?

The half-life of Trimelamol in human plasma at 37°C is 40.2 min., 15.5 min. in whole blood and 6.5 min. *in vivo* (39). Borm *et al* (46) reported studies with HMM and MHMPMM claiming to show that N-hydroxymethylmelamines could form a "protected" glutathione adduct which was not directly susceptible to chemical decomposition A number of observations were made which were claimed to support this hypothesis. Firstly that the production of PMM from HMM in intact rat hepatocytes was always less than the estimated loss of the parent compound HMM. Secondly, if reduced glutathione (GSH) levels were greatly reduced by pre-treating the animals with phorone, PMM production increased almost to the level predicted. It was claimed that no further metabolism was taking place thus the discrepancy in PMM production must be related to the presence or absence of GSH. It was also claimed that the half-life of MHMPMM was "minutes" in a liver 100,000g cytosolic preparation due to "enzymic degradation" but reverted to 105 min following denaturation. The means of denaturation were unspecified.

These claims were surprising in view of the known instability of N- hydroxymethyl-melaines and the hitherto similar behaviour of MHMPMM and Trimelamol in a wide variety of buffers or biological preparations (C. Rutty, personal communication). It was therefore decided to examine the role of liver cytosol and GSH in promoting the decomposition of Trimelamol.

Liver cytosol was prepared from Balb C mice using 50 mM phosphate buffer pH 7.4 as in the studies of Borm *et al*. GSH depletion was achieved by prior treatment with phorone 650 mg/kg 2 hours before the animals were killed (49). The liver cytosol was denatured by heating at 100°C for 10 min.

The stability of Trimelamol was studied at 37°C using an HPLC method used in pharmacokinetic studies (39). The addition of liver cytosol did result in a decrease in Trimelamol mean half-life from 12.75 min. in phosphate buffer alone to 7.75 min. This effect was largely abolished by prior treatment with phorone and totally abolished by heat denaturation.

It remains a possibility that direct enzymatic degradation is responsible for the enhanced decomposition of Trimelamol in liver cytosol. However, an alternative explanation is that formaldehyde dehydrogenase, a cytosolic enzyme which uses GSH and NAD^+ as cofactors, is influencing the breakdown by the oxidation of formaldehyde to formate leading to a mass action effect. The breakdown of Trimelamol with release of formaldehyde is a potentially reversible reaction, a fact which is utilised in the synthesis of the drug. In support of this hypothesis it was shown that oxidation of formaldehyde in the liver cytosol was decreased following phorone pre-treatment. The rate of formaldehyde oxidation *in vivo* in man is such that this effect could well be responsible for the short half-life and would also ensure that no appreciable toxicity resulted from the formaldehyde released during decomposition.

Mechanism of Action

Given the long history of this group of compounds it is perhaps surprising that the mechanism of action of HMM and its analogues such as Trimelamol remains unknown. Because of the structural similarity to triethylenemelamine, a known alkylating agent, it was proposed originally that HMM also worked by alkylation (18). However, HMM and its metabolites do not react with nitrobenzylpyridine (19) and HMM does not cause DNA-DNA cross-links even in the presence of an activating system (48). Furthermore there are reports of clinical activity with HMM in patients resistant to classical bifunctional alkylating agents (9,10).

Having identified the need for metabolic activation of HMM and PMM (22) attention was focused on the N-hydroxymethylmelamines. These were found to be much more cytotoxic *in vitro* than the 'parent' methylmelamines in the absence of an activating system (23). One such compound, MHMPMM, has been identified as a major metabolite of HMM *in vitro* (49) and *in vivo* in mice (50). MHMPMM is also capable of covalent binding to microsomal protein and calf thymus DNA (51).

If N-hydroxymethylmelamines are to be regarded as the active species resulting from oxidative N-demethylation of HMM and PMM, the role of formaldehyde, if any, needs to be considered. The former compounds are chemically unstable yielding formaldehyde on decomposition (18). The distinction between the non-selective cytotoxicity of formaldehyde and that of hydroxymethylmelamines was studied in detail by Rutty and Abel (23) using semicarbazide as a formaldehyde trapping agent. They demonstrated that semicarbazide penetrated readily into PC6 and L1210 cells *in vitro* but whereas pre-incubation with semicarbazide largely reversed the *in vitro* toxicity of MHMPMM and Trimelamol versus this resistant strain of the L1210 leukaemia, it had no effect on the toxicity to PC6 cells. Walker cells were intermediate in this regard. This observation has since been confirmed using a colony forming assay (B. Millar, personal communication). The significance of these observations lies in the fact that MHMPMM and Trimelamol are active against the PC6 tumour *in vivo* (26), marginally active versus the Walker tumour (38,39) but totally inactive against this particular strain of the L1210 (52), these findings suggest that free formaldehyde is not involved in the selective antitumour activity of these agents *in vivo* but may account for non-selective toxicity.

In addition to DNA binding (35,51) the type of DNA damage caused by N-hydroxymethylmelamines has been studied using the alkaline elution technique (48,52). Muindi found that formaldehyde produced DNA single-strand breaks and DNA-protein cross-links in the Chester Beatty strain of the L1210 leukaemia *in vitro*.[52] MHMPMM also produced DNA-protein cross-links and in the case of both drugs cross-linking was prevented by pre-incubation with semicarbazide. In this cell line semicarbazide also prevented cytotoxicity due to these agents (23). In contrast, the behaviour of PC6 cells was crucially different. Whilst formaldehyde again produced semicarbazide-reversible DNA-protein cross-links, no cross-links were observed with MHMPMM. Furthermore the cytotoxic effect of this drug and that of Trimelamol was not reversed by semicarbazide, although this did prevent formaldehyde toxicity (23). The conclusion appeared to be that formaldehyde release was responsible for the cytotoxicity and DNA-protein cross-linking observed when L1210 cells were treated with MHMPMM. However, in PC6 cells free formaldehyde is not available to react with semicarbazide, hence the lack of protection and DNA-protein cross-links. There could be several explanations for this observation but at present the difference between cell types is unexplained. What is clear is that these data support the view that formaldehyde release does not significantly contribute to the selective toxicity of MHMPMM and Trimelamol versus this particular sensitive cell line.

Ross *et al* reported slightly different results using a partially sensitive strain of the L1210 leukaemia (48). This behaved rather like the PC6 tumour in that semicarbazide pre-treatment did not prevent cytotoxicity due to activated HMM or PMM, nor due to MHMPMM or Trimelamol. Formaldehyde produced extensive DNA-protein cross-links but not strand breaks in this cell line, but MHMPMM had little effect. Activated HMM and PMM produced a small degree of DNA-protein cross-linking. Surprisingly, Trimelamol was said to produce DNA-DNA cross-links. This raises the possibility that a polyfunctional agent like Trimelamol could be acting by a different mechanism. However, MHMPMM and Trimelamol have similar activity *in vitro* versus the PC6 tumour (23) and the NCI L1210 (48). Therefore it seems unlikely that cross-linking behaviour is crucial to the mechanism of action of these agents although this issue remains unresolved.

A number of experiments have been performed with Trimelamol to try and determine the importance of interactions with DNA. It has been demonstrated that Trimelamol does not inhibit the incorporation of thymidine into calf thymus DNA. As with MHMPMM, semicarbazide is capable of abolishing the inhibitory effect of Trimelamol on thymidine incorporation into PC6 cells *in vitro*, but it has little effect on inhibition of growth by Trimelamol (53). Similar experiments were performed in which Trimelamol was administered at a dose of 50 mg/kg IV daily x 3 to mice bearing the ascitic form of the PC6 tumour. This dose resulted in 77% prolongation of survival confirming that a therapeutically relevant dose had been chosen and that the tumour retained its sensitivity to Trimelamol when passaged in this way. Tumour cells were subsequently harvested and DNA synthesis measured *in vitro* by thymidine incorporation. Thus it was hoped that non-specific effects of formaldehyde would be excluded. A small reduction of 29% in thymidine incorporation was observed 2 hours after drug administration, however by 5 hours this had returned to 95% of the control value. Inhibition of DNA synthesis does not appear to be important *in vivo*. Clearly further work is required to investigate the precise nature of any DNA adducts formed by Trimelamol or identify an alternative target for this drug.

Clearly Trimelamol represents an intriguing development, in that it demonstrates that the administration of a HMM analogue which does not require metabolic activation can produce predictable cytotoxic activity. The difficulties with formulation and the combination of a short elimination half-life with the need for repeated or prolonged exposure are serious problems and it is possible that a more stable and soluble analogue would be advantagous. The activity in 'platinum-resistant' ovarian cancer is interesting and provides a stimulus for further research. If it could be shown that N-methyl-melamines do indeed have a unique mechanism of action then rational design of alternative structures might be feasible. Therefore, in conclusion the future of this class of compounds appears to depend on the identification of their mechanism of action and the subsequent development of Trimelamol analogues with superior solubility, stability and perhaps anti-tumour activity.

REFERENCES

1.   D.W. Kaiser, J.T. Thurston, J.R. Dudley, F.C. Schaefer, I. Hechenbleikner and D. Holm-Harsen, Cyanuric chloride derivatives II, J. Am. Chem. Soc. 73:2984 (1951).

2.   J.A. Hendry, F.L. Rose, A.L. Walpole, Cytotoxic agents III. Derivatives of ethyleneimines, Br. J. Pharmacol. 6:357 (1951)

3.   J.A. Hendry, F.L. Rose, A.L. Walpole, Cytotoxic agents: I, methylolamides with tumour inhibitory activity, and related inactive compounds. Br. J. Pharmacol. 6:201 (1951).

4.   L.W. Wilson and J.G. de la Garza, Phase I study of hexamethylmelamine (NSC-13875), Cancer Chemother. Rep. 48: 49-53 (1965).

5.   J. Louis, N.B. Louis, J.W. Linman, W.J. Donnelly, B.L. Isaacs and S.O. Schwartz, The clinical pharmacology of hexamethylmelamine: Phase I study, Clin. Pharm and Therapeutics 8:55 (1967).

6.   R.H. Blum, R.B. Livingston, S.K. Carter, Hexamethylmelamine - a new drug with activity in solid tumours. Europ. J. Cancer 9:195-202 (1973).

7.   Legha SS, Slavik M, Carter SK (1976) Hexamethylmelamine and evaluation of its role in the therapy of cancer. Cancer 38:27-35

8.   R.I. Fisher, R.C. Young, Advances in the staging and treatment of ovarian cancer. Cancer 39:967-972 (1977).

9.   B.L. Johnson, R.I. Fisher, R.A.Bender, V.T. DeVita, B.A. Chabner, R.C Young, Hexamethylmelamine in alkylating agent-resistant ovarian carcinoma. Cancer 42:2157-2161 (1978).

10.   P.D. Bonomi, J. Mladineo, B. Morrin, G. Wilbanks, R.E. Clayton, Phase II trial of hexamethylmelamine in ovarian carcinoma resistant to alkylating agents. Cancer Treat. Rep. 63:167 (1979).

11.   R.C. Young, B.A. Chabner, S.P. Hubbard, R.I. Fisher, R.A.Bender, T. Anderson, R.H. Simon, G.P. Canellos, V.T. DeVita, Advanced ovarian adenocarcinoma: a prospective clinical trial of melphalan (L-PAM) versus combination chemotherapy. New Eng. J. Med. 229:1261-1266 (1978).

12.   S.E. Vogl, M. Berenzweig, B.H. Kaplan, M. Moukhtar, W. Buskin, The CHAD and HAD regimens in advanced ovarian cancer: combination chemotherapy including cyclophosphamide, hexamethylmelamine, adriamycin and cis-dichlorodiammineplatinum(II). Cancer Treat. Rep. 63:311-317 (1979).

13.   J.P. Neijt, W.W. ten Bockel Huinink, M.E.L. van der Burg, A.T. van Oosterom, C.D. Kooyman, J.C. van Houwelingen, H.M. Pinedo, Combination chemotherapy with or without hexamethylmelamine in alkylating agent resistant ovarian carcinoma. Cancer 53:1467-1472 (1984).

14.   G.A. Omura, C.P. Morrow, J.A. Blessing, A. Miller, H.J. Buchsbaum, H.D. Homesley, L. Leone, A randomised comparison of melphalan versus melphalan plus hexamethylmelamine versus adriamycin plus cyclophosphamide in ovarian carcinoma. Cancer 51:783-789 (1983).

15.   B.J. Foster, K. Clagett-Carr, S. Marsoni, R. Simon and B. Leyland-Jones, Role of hexamethylmelamine in the treatment of ovarian cancer: Where is the needle in the haystack? Cancer Treat. Rep. 70:1003-1014 (1986).

16.   H.W. Bruckner, C.J. Cohen, E. Feuer, J.F. Holland, B. Kabakow, R. Wallach, Long-term follow-up of stage III and IV ovarian cancer: controlled clinical trials utilising cisplatin (P), doxorubicin (A), cyclophosphamide (C) and hexamethylmelamine (H). Proc. Am. Soc. Clin. Oncol. 6:121 (1987).

17.   F.B. Stehman, C.E. Ehrlich, R.N. Callangan, Failure of hexamethylmelamine as salvage therapy in ovarian epithelial adenocarcinoma resistant to combination chemotherapy. Gynecol. Oncol. 17:189-195 (1984).

18.   A.B. Borkovec and A.B. De Milo, Insect chemosterilants V, derivatives of melamine, J. Med. Chem. 10:457-461 (1967).

19.   J.F. Worzalla, B.M. Johnson, G. Ramirez, G.T. Bryan, N-demethylation of the antineoplastic agent hexamethylmelamine by rats and man. Cancer Res. 33:2810-2815 (1973).

20.   J.F. Worzalla, B.D. Kaiman, B.M. Johnson, G. Ramirez, G.T. Bryan, Metabolism of hexamethylmelamine ring-^{14}C in rats and man. Cancer Res. 34:2669-2674 (1974).

21.   J.F. Worzalla, D.M. Lee, R.O. Johnson, G.T. Bryan, Effects of microsomal enzyme-inducing chemicals on the metabolism of hexamethylmelamine (HMM, NSC-13875) in rats and hamsters. Proc. Am. Ass. Cancer Res. 12:41 (1972).

22.   C.J. Rutty and T.A. Connors, In vitro studies with hexamethylmelamine. Biochem. Pharmacol. 26:2385-2391 (1977).

23.   C.J. Rutty, G. Abel, In vitro cytotoxicity of the methylmelamines. Chem. Biol. Interactions 29:235-246 (1980).

24.   R.L. Richardson, M.M. Ames, J.S. Kovach, C.G. Moertel, Clinical evaluation of a new formulation of hexamethylmelamine (H) suitable for intravenous administration. Proc. Am. Ass. Cancer Res. 27:662 (1986).

25.   M. D'Incalci, G. Bolis, C. Mangioni, L. Morasca, S. Garattini, Variable oral absorption of hexamethylmelamine in man. Cancer Treat. Rep. 62:2117-2119 (1978).

26.   A.J. Cumber and W.C.J. Ross, Analogues of hexamethylmelamine. The anti-neoplastic activity of derivatives with enhanced water solubility. Chem. Biol. Interactions 17:349-357 (1977).

27.   T.A Connors, A.J. Cumber, W.C.J. Ross, S.A. Clarke, B.C.V. Mitchley, Regression of human lung tumor xenografts induced by water-soluble analogs of hexamethylmelamine. Cancer Treat. Rep. 61:927-928 (1977).

28.   E.S. Casper, R.J. Gralla, R.L. Garrett, B.R. Jones, T.M. Woodcock, C. Gordon, D.P. Kelsen, C.W. Young, Phase I and pharmacological studies of pentamethylmelamine administered by a 24-hour intravenous infusion. Cancer Res. 41:1402 (1981).

29.   R.S. Goldberg, J.P. Griffin, J.W. McSherry and I.H. Krakoff, Phase I study of pentamethylmelamine, Cancer Treat. Rep. 64:1319 (1980).

30.   D.C. Ihde, J.S. Dutcher, R.C. Young RC, R.L. Cordes, A.L. Barlock, S.M.Hubbard, R.B. Jones, M.R. Boyd, Phase I trial of pentamethylmelamine: a clinical and pharmacologic study. Cancer Treat. Rep. 65:755 (1981).

31.   J.R.F. Muindi, D.R. Newell, I.E. Smith, K.R. Harrap, Pentamethylmelamine (PMM): Phase I clinical and pharmacokinetic studies. Br. J. Cancer 47:27 (1983).

32.   D.A. Van Echo, D.F. Chiuten, M. Whitacre, J. Aisner, J.L. Lichtenfeld, P.H. Wiernik, Phase I trial of pentamethylmelamine in patients with previously treated malignancies. Cancer Treat. Rep. 64:1335 (1980).

33.   C.J. Rutty, D.R. Newell, J.R.F. Muindi, K.R. Harrap, The comparative pharmacokinetics of pentamethylmelamine in man, rat and mouse. Cancer Chemother. Pharmacol. 8:105-111 (1982).

34.   C.J. Rutty, PhD Thesis, University of London (1978).

35.   M. Morimoto, D. Green, A. Rahman, A. Goldin, P.S. Schein, Comparative pharmacology of pentamethylmelamine and hexamethylmelamine in mice. Cancer Res. 40:2762-2767 (1980).

36.   Screening Data Summary, Drug Evaluation Branch, Division of Cancer Treatment, National Cancer Institute, Maryland, USA (1977).

37.   E. Boven, N.M. Nauta, H.M.M. Schluper, C.A.M. Erkelens, H.M. Pinedo, Superior efficacy of trimelamol to hexamethylmelamine in human ovarian cancer xenografts. Cancer Chemother. Pharmacol. 18:124-128 (1986).

38.   C.J. Rutty, D.R. Newell, J.R.F. Muindi, K.R. Harrap, Development of potential clinical alternatives to hexamethylmelamine. The Control of Tumour Growth and Its Biological Basis, Davis W, Maltoni C, Tenneberger St. (eds), Academie Verlag, Berlin pp180-188 (1983).

39. C.J. Rutty, I.R. Judson, G. Abel, P.M. Goddard, D.R. Newell, K.R. Harrap, Preclinical toxicology, pharmacokinetics and formulation of N^2,N^4,N^6-trihydroxymethyl-N^2,N^4,N^6-trimethylmelamine (Trimelamol) a water-soluble cytotoxic s-triazine which does not require metabolic activation. Cancer Chemother. Pharmacol. 17:251-258 (1986).

40. D.R. Newell, C.J. Rutty, J.R.F. Muindi, K.R. Harrap, Experimental studies on trimethyl-trimethylolmelamine as an alternative to hexamethylmelamine (HMM) and pentamethylmelamine (PMM). Br. J. Cancer 44:281 (1981).

41. I.R. Judson, C.J. Rutty, G. Abel, M.A. Graham, Low central nervous system penetration of N^2,N^4,N^6-trihydroxymethyl-N^2,N^4,N^6-trimethylmelamine (Trimelamol): A cytotoxic s-triazine with reduced neurotoxicity. Br. J. Cancer 53:601-606 (1986).

42. I.R. Judson, A.H. Calvert, C.J. Rutty, G. Abel, L.A. Gumbrell, M.A. Graham, B.D. Evans, D.E.V. Wilman, S.E. Ashley and F. Cairnduff, Phase I trial and pharmacokinetics of Trimelamol (N^2,N^4,N^6-Trihydroxymethyl-N^2,N^4,N^6-trimethylmelamine), Cancer Res. 49:5475-5479 (1989).

43. J.M. Collins, D.S. Zaharko, R.L. Dedrick and B.A. Chabner, Potential roles for preclinical pharmacology in phase I clinical trials. Cancer Treat. Rep. 70:73-80 (1986).

44. EORTC Pharmacokinetics and Metabolism Group, Pharmacokinetically guided dose escalation in phase I clinical trials: commentary and proposed guidelines. Europ. J. Cancer Clin. Oncol. 23:1083-1087 (1987).

45. I.R. Judson, M.E. Gore, L.A. Gumbrell, K. Balmanno, D.I. Jodrell, T.J. Perren, E. Wiltshaw, P. Blake and A.H. Calvert, A phase II evaluation of Trimelamol (N^2,N^4,N^6-trihydroxymethyl-N^2,N^4,N^6-trimethylmelamine)in stage III/IV ovarian cancer, Br.J. Cancer, 58:273 (1988).

46. P.J.A. Borm, M.J.J. Mingels, C. Ank, F. Sierevogel, M. van Graft, A. Hulshoff and J. Noordhoek, Cellular and subcellular studies of the biotransformation of hexamethylmelamine in rat isolated hepatocytes and intestinal epithelial cells, Cancer Res. 44:2820-2826 (1984).

47. R. van Doorn, CH-M Leijdekkers and P.TH. Henderson, Synergistic effects of phorone on the hepatotoxicity of bromobenzene and paracetamol in mice, Toxicology 11:225-233 (1978).

48. W.E. Ross, D.R. McMillan and C.F. Ross, Comparison of DNA damage by methylmelamines and formaldehyde, J. Natl.Cancer Inst. 67:217-221 (1981).

49. W.E. Ross, D.R. McMillan, C.F. Ross, Comparison of DNA damage by methylmelamines and formaldehyde. JNCI 67:217-221 (1981).

50. A. Gescher, M. D'Incalci, R. Fanelli, P. Farina, N-hydroxymethylpentamethylmelamine a major in vitro metabolite of hexamethylmelamine. Life Sci. 26:147-154 (1980).

51. J. Dubois, G. Atassi, M. Hanocq, F. Abikhalil, Pharmacokinetics and metabolism of hexamethylmelamine in mice after IP administration. Cancer Chemother. Pharmacol. 18:226-230 (1986).

52. M.M. Ames, M.E. Sanders, W.S. Tiede, Role of N-methylolpentamethylmelamine in the metabolic activation of hexamethylmelamine. Cancer Res. 43:500-504 (1983).

53. J. Muindi, PhD Thesis, London University, (1981).

54. I.R. Judson, MD Thesis, The Development of the new anti-cancer drug Trimelamol, Cambridge University (1989).

EXPERIMENTAL BACKGROUND AND EARLY CLINICAL STUDIES
WITH IMIDAZOTETRAZINE DERIVATIVES

E.S. Newlands[1], G.R.P. Blackledge[2], J. Slack[3]
N.S.A. Stuart[2], and M.F.G. Stevens[3]

[1]Dept. of Medical Oncology, Charing Cross Hospital
London W6 8RF
[2]Dept. of Medicine, Queen Elizabth Hospital
Birmingham B15 2TH
[3]Dept. of Pharmaceutical Sciences, Aston University
Birmingham B4 7ET

Introduction

A series of imidazotetrazine derivatives were synthesised by Stevens and colleagues in the late 70's (1). The activity of one of these derivatives, mitozolamide (8-carbamoyl-3-(2-chloro-ethyl)imidazo[5-1-d]-1,2,3,5-tetrazin-4(3H)-one) was identified and the compound's ability to cross link DNA confirmed (2, 3, 4).

Further studies identified the broad spectrum antitumour activity of mitozolomide and that its degradation to the acyclic monochloroethyl triazine (MCTIC) (Fig. 1) was probably the active product of mitozolomide (5). From the various analogues of mitozolomide that were synthesised the most interesting was the methyl analogue, temozolomide (CCRG 81045; M & B 39831; NSC 362856). This analogue was of interest since it could readily ring open to the linear triazine MTIC (Fig. 2). MTIC is thought to be the active metabolite of dacarbazine and the potential advantage of temozolomide was the spontaneous ring opening to the active metabolite in vivo (6). One reason why dacarbazine is thought to be of very limited activity in man is its poor metabolic activation.

Mechanism of action

Early studies suggested that mitozolomide was cytotoxic through the production of MCTIC (7). The major site of action of mitozolomide is thought to be the alkylation of the O^6 position of guanine (8). In addition to this site of action, mitozolomide also alkylates guanine at the N7 position (9).

Resistance to mitozolomide is associated with the capacity of the repair of the alkylation at the O^6 position of guanine (10). This indicates the importance of damage at this site in inducing cytotoxicity. Comparison of mitozolomide with temozolomide in their relative ability to alkylate DNA showed differences both with isolated DNA and in intact cells (11). An

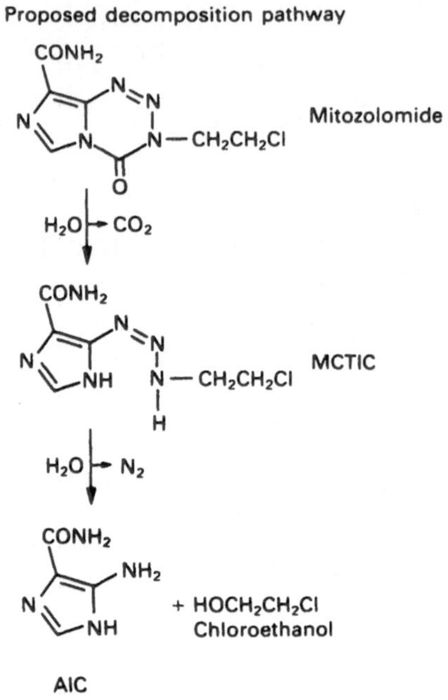

Proposed decomposition pathway

Figure 1

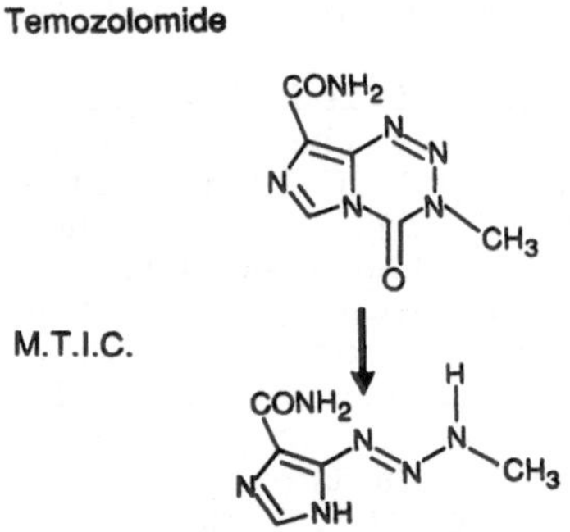

Figure 2

interesting recent study in haemopoietic stem cells has shown that when these are transfected to express E.Coli DNA alkyltransferase, a normally sensitive haemopoietic stem line acquires resistance to mitozolomide (12).

Experimental activity

Mitozolomide has a very broad spectrum antitumour activity against murine systems (5, 13). Significant activity of mitozolomide was also seen in the following human tumour xenografts: melanomas, sarcomas, lung and colon carcinomas (14).

In retrosepct this dramatic antitumour activity requires closer analysis. It is well recognised that ultimately the biochemistry of murine tumours and human tumours is different since a range of cytotoxic agents which are curative in murine systems show only limited activity in the clinic. This is particularly so of the nitrosoureas and nitrosourea-like drugs such as the imidazotriazines. Using a different murine system, the MAC system (15) showed that mitozolomide only had activity near the toxic dose, which is in fact what has been found in the clinic.

Temozolomide also showed broad spectrum activity in murine systems (Tables 1 & 2).

The imidazotriazines have a number of properties that are attractive from the point of view of the clinic. They show good bioavailability and tissue distribution. In addition, both mitozolomide and temozolomide have good penetration into the central nervous system (16). Some schedule dependency has been shown with the imidazotriazines and a daily x 5 days schedule with temozolomide showed increased therapeutic activity against both P388 and L1210 leukaemias (6).

Mitozolomide Phase I and II studies

Mitozolomide entered clinical trials in 1983, initially with an intravenous formulation in DMSO. The good bioavailability of mitozolomide was confirmed (17). The doses studied ranged from 8 mg/m^2 to 153 mg/m^2. Overall the symptomatic tolerance to mitozolomide was good with mild to moderate nausea and vomiting. Little myelosuppression was seen in doses up to 115 mg/m^2 in this study but severe myelosuppression was found in doses above this. The dose-limiting toxicity was thrombocytopaenia which was profound and recovery was delayed in some patients for up to 6-8 weeks. Subsequent experience with mitozolomide showed that the initially recommended dose of 115 mg/m^2 was still myelosuppressive despite little evidence of this being found in the initial eight patients treated at this dose. Most Phase II studies were subsequently performed at 90 mg/m^2 or even as low as 70 mg/m^2 given once every six weeks. In the initial Phase I study there were hints of clinical activity in ovarian adenocarcinoma.

A number of Phase II studies were performed with mitozolomide in the following diseases: colorectal and breast carcinoma (18), bladder cancer (19), ovarian adenocarcinoma (20, 21) and renal cell carcinoma (22), and all these Phase II studies were negative with no responses being seen and varying degrees of myelosuppression being dose-limiting.

However, not all the Phase II studies with mitozolomide showed no evidence of activity. Some activity was seen in malignant melanoma (23). In addition, some activity was seen in small cell carcinoma of the lung. In a summary of the clinical data to May 1986, 221 patients had been treated with 146 available for assessment of response, giving a total of 2 complete responses and 6 partial responses (all these responses being seen in malignant melanoma and small cell carcinoma of the lung).

Table 1

Tumour	Temozolomide	Dacarbazine
	(50 mg/kg days 1,2,3,4)	
B16 melanoma s.c.	++	+
C26 adenocarcinoma s.c.	+	-
L1210 leukaemia i.p.	+	+
L1210 leukaemia i.v.	++	-
Lewis lung i.m.	++	++
M5076 carcinoma s.c.	++	++
P388 leukaemia i.p.	-	-
P388 leukaemia i.v.	-	-
Rat Walker carcinoma	++	+
(days 1,2,3,4,7)	(12.5)	(50)

+ = greater than 25% increase in life span but less than 50% increase for L1210 and P388.

++ = greater than 50% increase in life span for L1210 and P388.

+ = greater than 30% tumour weight inhibition but less than 80% for B16, C26, M5076, Lewis, Walker.

++ = greater than 80% tumour weight inhibition for B16, C26, M5076, Lewis, Walker.

Table 2

Temozolomide Experimental Antitumour Activity II

G16	Adenocarcinoma
G26	Adenocarcinoma
MX-1	Xenograft
M5076	Carcinoma
PC-6	Plasmacytoma
TLX-5	Lymphoma

Mitozolomide has also been studied on a schedule given over five consecutive days. Again on this schedule the dose-limiting toxicity was thrombocytopaenia at 15 mg/m² per day x 5 (24). Given the limited clinical activity of mitozolomide, this schedule has not been extensively studied.

Temozolomide Phase I

Since the Phase I and Phase II studies of mitozolomide had shown limited clinical activity but unpredictable and severe dose-limiting toxicity and in particular thrombocytopaenia, it was decided to study the analogue, temozolomide, in a Phase I study. This was started in 1987, again with the initial formulation in DMSO being intravenous in a dose of 50 mg/m². Temozolomide is generally well tolerated with mild to moderate nausea and vomiting at higher doses. The initial pharmacokinetic data of temozolomide at doses between 50 mg/m² and 200 mg/m² are shown in Table 3. At 200 mg/m² the good bioavailability of

Table 3

TEMOZOLOMIDE PHARMACOKINETICS - SUMMARY

MODE	DOSE mg/m^2	$t_{1/2}$ h	AUC (0-τ) h.ug/ml	V_d l	Cl_T ml/min	F
i.v.	50	1.23	4.89	30.8	289.7	
i.v.	50	1.78	8.79	27.2	176.4	
i.v.	100	1.58	13.8	31.3	229.4	
i.v.	100	1.56	20.8	19.0	141.4	
i.v.	150	1.62	30.7	14.3	101.9	
i.v.	150	1.56	26.6	15.8	117.5	
i.v.	200	0.90	36.6	11.4	145.8	
oral	200	1.61	42.9	17.3	124.4	1.17
i.v.	200	1.71	46.2	19.2	130.0	
oral	200	1.71	42.3	21.0	141.9	0.92
i.v.	200	1.52	30.2	27.6	209.7	
i.v.	200	1.06	34.5	12.4	135.4	
oral	200	1.39	32.3	17.4	144.5	0.94

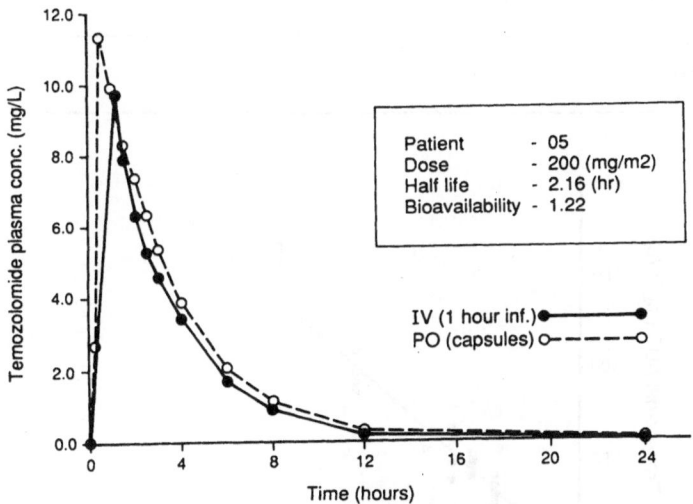

Figure 3 **Temozolomide Bioavailability**

Table 4   Phase I Trial of Temozolomide

Single dose schedule:

	Dose (mg/m²)	No. of patients entered	No. of doses administered	Evaluable kinetics
	50	3	5	2
	100	4	11	2
	150	3	5	0
i.v.	200	8	10	4
p.o.	200	9	9	7
"	250	4	5	3
"	300	5	8	5
"	360	3	3	1
"	430	1	1	0
"	520	7	9	4
"	700	7	14	9
"	750	2	2	1
"	920	6	6	1
"	900	7	10	5
"	1000	6	9	0
"	1200	1	1	0
Totals		49	105	44

150.0 mg/m² daily x 5 schedule

Dose (mg/m²) Daily x 5		
150	9	17

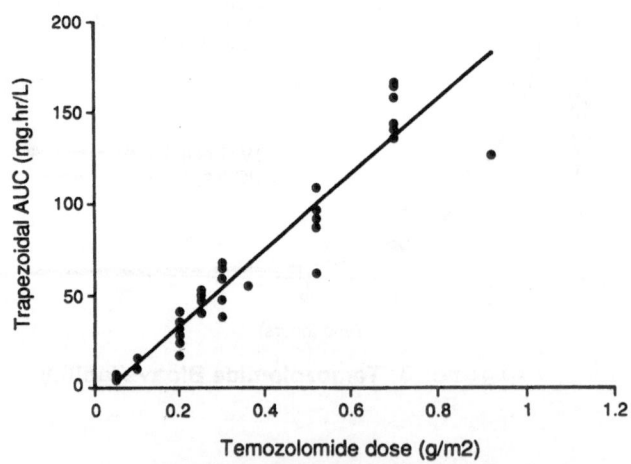

Temozolomide Phase I Trial

Figure 4

temozolomide was confirmed (Fig. 3). The pharmacokinetics of temozolomide have been performed at all doses and there is a linear correlation between the AUC and the dose administered (Fig. 4). Subsequent dose escalation up to 1,200 mg/m² (Table 4) has shown that the dose-limiting toxicity with temozolomide is again somewhat unpredictable myelosuppression with thrombocytopaenia being dose-limiting. On the single dose schedule, hints of clinical activity were seen in malignant melanoma and in one patient with squamous cell carcinoma of the head and neck (25). In view of the schedule dependency of temozolomide (6), the Phase I of temozolomide is continuing on a daily x 5 schedule. This schedule is fairly well tolerated with WHO grade 2 - 3 nausea and vomiting in most patients and this is usually well controlled with antiemetics. Further hints of clinical activity have been seen at a dose of 150 mg/m² p.o. daily x 5 in patients with malignant melanoma with one complete response in cutaneous metastases (Table 4). Curiously no thrombocytopaenia has been seen in the initial first nine patients treated on this schedule. However, from the now quite extensive clinical experience of the imidazotriazines, one has to be cautious about the rather unpredictable myelotoxicity. At present it is proposed to go up to the next dose level of 180 mg/m² daily p.o. x 5 schedule which is likely to be near the MTD. It is proposed to go ahead with a Phase II study in malignant melanoma, given the drug's good tolerance and initial signs of clinical activity in this disease.

Summary

Some of the imidazotriazines have shown dramatic preclinical activity against murine tumours and human tumour xenografts. However, the lead compound, mitozolomide, has shown unpredictable major and prolonged myelosuppression at doses producing minimal clinical activity although hints of activity were seen in malignant melanoma and small cell carcinoma of the lung. Temozolomide shows many similar characteristics in the clinic to mitozolomide, again with myelosuppression being dose-limiting. However, the initial clinical experience suggests that this toxicity is somewhat more predictable than with mitozolomide and it may be worthwhile pursuing temozolomide in a limited number of Phase II studies.

REFERENCES

1. M.F.G. Stevens. The medical chemistry of 1,2,3 triazines. Prog. Med. Chem. 13 : 205-269 (1976).
2. K. Vaughan and M.F.G. Stevens. Monoalkyltriazines. Chem. Soc. Rev. 7 : 377-397 (1978).
3. M.F.G. Stevens, J.A. Hickman, R. Stone et al. Antitumour imidazotetrazines 1. Synthesis and chemistry of 8-carbamoyl 3-(2-chloroethyl)imidazo[5,1,D]-1,2,3,5-tetrazin-4(3H)-one, a novel broad spectrum antitumour agent. J. Med. Chem. 27: 196-201, 1984.
4. N.W. Gibson, L.C. Erickson & J.A. Hickman. Effects of the antitumour agent 8-carbamoyl-3-(2-chloroethyl)imidazo[5,1-d]-1,2,3,5-tetrazin-4(3H)-one on the DNA of mouse L1210 cells. Cancer Res. 44 : 1767-1771 (1984).
5. N.W. Gibson, J.A. Hickman & L.C. Erickson. DNA cross-linking and cytotoxicity in normal and transformed human cells treated in vitro with 8-carbamoyl-3-(2-chloroethyl) imidazo[5,1-d]-1,2,3,5-tetrazin-4(3H)-one. Cancer Res. 44 : 1772-1775 (1984).

6. M.F.G. Stevens, J.A. Hickman, S.P. Langdon, D. Chubb, L. Vickers, R. Stone, G. Baig, C. Goddard, N.W. Gibson, J.A. Slack, C. Newton, E. Lunt, C. Fizames and & F. Lavelle. Antitumour activity and pharmacokinetics in mice of 8-carbamoyl-3-methylimidazo[5,1-d]-1,2,3,5-tetrazin-4(3H)-one (CCRG 81045; M & B 39831), a novel drug with potential as an alternative to dacarbazine. Cancer Res. 47 : 5846-5852 (1987).

7. C.M. Horgan & M.J. Tisdale. Antitumour imidazotetrazines - IV. An investigation into the mechanism of antitumour activity of a novel and potent antitumour agent, mitozolomide (CCRG 81010; M & B 39565; NSC 353451). Biochem. Pharmacol. 33 (14) : 2185-2192 (1984).

8. N.W. Gibson, J.A. Hartley, W.B. Mattes, K.W. Kohn, L.C. Erickson. The effects of pretreatment of human tumor cells with MNNG on the DNA crosslinking and cytotoxicity of two chloroethylating agents. Abstract: International Conference on Mechanisms of DNA Damage and Repair. Implications for Carinogenesis and Risk Assessment. Gaithersburg, Maryland. (June 2 -7, 1985).

9. J.A. Hartley, N.W. Gibson, K.W. Kohn, W.B. Mattes. DNA sequence selectivity of guanine-N7 alkylation by three antitumor chloroethylating agents. Cancer. Res. 46 (4, Pt. 2) : 1943-1947 (1986).

10. M.J. Tisdale. Antitumour imidazotetrazines - XV. Role of guanine O (6) alkylation in the mechanisms of cytotoxicity of imidazotetrazines. Biochem. Pharmacol. 38 (4) : 457-462 (1987).

11. V.L. Bull, M.J. Tisdale. Antitumour imidazotetrazines-XVI. Macromolecular alkylation by 3-substituted imidazo-tetrazines. Biochem. Pharmacol. 36 (19) : 3215-3220 (1987).

12. J. Jelinek, K. Kleibl, T.M. Dexter & G.P. Margison. Increased resistance to the toxic effects of mono and bifunctional alkylating agents in murine hemopoietic cells expressing DNA alkyltransferase genes from E. coli. Proc. Ann. Meeting, Amer. Assoc. Cancer Res. 28 : 112. Abstract (1987).

13. J.A. Hickman, M.F. Stevens, N.W. Gibson, S.P. Langdon, C. Fitzjames, F. Lavelle, G. Atassi, E. Lunt & R.M. Tilson. Experimental antitumour activity against murine tumor model systems of 8-carbamoyl-3-(2-chloroethyl)imidazo[5,1,d]-1,2, 3,5-tetrazin-4(3H)-one(mitozolomide), a novel broad spectrum agent. Cancer. Res. 45 (7) : 3008-3013 (1985).

14. O. Fodstad, S. Aamdal, A. Pihl & M.R. Boyd. Activity of mitozolomide (NSC 353451), a new imidazotetrazine, against xenografts from human melanomas, sarcomas, and lung and colon carcinomas. Cancer Res. 45, 1778-1786 (1985).

15. M.C. Bibby, J.A. Double, I.A. Wahed, N. Hirbawi & T.G. Baker. The logistics of broader pre-clinical evaluation of potential anti-cancer agents with reference to anti-tumour activity and toxicity of mitozolomide. Br. J. Cancer 58 (2) : 139-143 (1988).

16. C.J. Brindley, P. Antoniw & E.S. Newlands. Plasma and tissue disposition of mitozolomide in mice. Br. J. Cancer 53, 91-97 (1986).

17. E.S. Newlands, G. Blackledge, J.A. Slack, C. Goddard, C.J. Brindley, L. Holden & M.F.G. Stevens. Phase I clinical trial of mitozolomide. Cancer Treatment Reports 69, 7-8 (July/Aug. 1985).

18. P. Herait, P. Rougier, J. Oliveira, P.M. Delgado, L.F. May, M. Hayat & J.P. Armand. Phase II study of mitozolomide (M & B 39565) in colorectal and breast cancer. Invest. New Drugs 6 (4) : 323-325 (1988).
19. G. Blackledge, J.T. Roberts, S. Kaye, R. Taylor, J. Williams, B. De Stavola & B. Uscinska. A phase II study of mitozolomide in metastatic transitional cell carcinoma of the bladder. Eur. J. Cancer. Clin. Oncol, 25/2 : 391-392 (1989).
20. M. Harding, D.Northcott, J. Smyth, N.S.A. Stuart, J.A. Green & E. Newlands. Short communication in: Phase II evaluation of mitozolomide in ovarian cancer. Br. J. Cancer 57 : 113-114 (1988).
21. J.P. Neijt, M.E. van der Burg, J.P. Guastalla, M. George, J.B. Vermorken & N. Rotmentsz. Mitozolomide in patients with advanced ovarian carcinoma: A Phase II study of the EORTC Gynecological Cancer Cooperative Group. Abstract: Proc. ECCO-4. Fourth European Conference on Clinical Oncology and Cancer Nursing; Federation of European Cancer Societies, Madrid. p. 214. (Nov. 1-4, 1987).
22. A.T. van Oosterom, G. Stoter, A.V. Bono, T.A.W. Splinter, S.D. Fossa, A.J. Verbaeys, P.H.M. de Mulder, M. de Pauw & R. Sylvester. Mitozolomide in advanced renal cancer: A Phase II study in previously untreated patients from the EORTC Genitourinary Tract Cancer Cooperative Group. Eur. J. Cancer Clin. Oncol. 25, 8 : 1249-1250 (1989).
23. S. Gundersen, S. Aamdal & O. Fodstad. Mitozolomide (NSC 353451), a new active drug in the treatment of malignant melanoma. Phase II trial in patients with advanced disease. Br. J. Cancer 55 : 433-435.
24. J.H. Schornagel, G. Simonetti & J.G. McVie. Phase I study of mitozolomide (NSC 353451) using a daily x 5 schedule. Proc. Fifth NCI-EORTC Symposium on New Drugs in Cancer Therapy. Abstract. (1986).
25. E.S. Newlands, J. Slack, G. Blackledge, N. Stuart, C. Quartermain, R. Hoffman & M.F.G. Stevens. Proc. Sixth NCI-EORTC Symposium on New Drugs in Cancer Therapy. Abstract No. 419. (1989).

O^6-ALKYLGUANINE–DNA–ALKYLTRANSFERASE: SIGNIFICANCE, METHODS OF MEASUREMENT AND SOME HUMAN TUMOUR AND NORMAL TISSUE LEVELS

(Report of a Mini-workshop)

G.P. Margison, P.J. O'Connor, D.P Cooper and J. Davies

Department of Carcinogenesis,
Paterson Institute for Cancer Research,
Christie Hospital, Manchester, M20 9BX   UK

C.N. Hall

Department of General Surgery
Wythenshawe Hospital, Manchester UK

S.M.S. Redmond, K. Buser and T. Cerny

Institut Fur Tumorforschung,
Tiefenauspital, 3004 Bern, Switzerland

L. Citti

Istituto Di Mutagenesi e Differenziamento CNR,
via Svezia 10, Pisa 56100, Italy

M. D'Incalci

Istituto "Mario Negri" 20157 Milano, Italy

SIGNIFICANCE OF ALKYLTRANSFERASE IN CARCINOGENESIS

There is a considerable amount of experimental evidence to indicate that O^6-alkylation of guanine in DNA is one of the principal factors responsible for the carcinogenic effect of alkylating agents. This has been reviewed extensively elsewhere (1-9) but can be briefly summarised as follows: O^6-alkylguanine miscodes during DNA replication and has been shown to be mutagenic; alkylating agents that can be considered as strong mutagens or potent carcinogens produce relatively higher levels of O^6-alkylguanine in tissue DNA than the less potent agents; in general, the extent of formation, persistence or accumulation (during chronic treatment schedules) of O^6-alkylguanine in DNA is greater in target tissues than in non-target tissues. Although formation and accumulation can be dependent on tissue-specific metabolic activation of an inactive agent, persistence and hence accumulation reflects tissue and cellular levels of the repair enzyme

O^6-alkylguanine-DNA-alkyltransferase (ATase). The existence of this enzyme in both pro- and eukaryotes itself implies that the lesion is in some way detrimental to the host. Thus mammalian cells that express relatively high levels of ATase, encoded by either endogenous or transfected genes are generally more resistant to the toxic, mutagenic and other biological effects of alkylating agents than those that express low levels of the enzyme (See refs 9,10).

The mechanism of malignant transformation by alkylating agents may, at least in some cases, be related to the activation of the H-ras oncogene. Following alkylation of the second guanine residue in codon 12, GC-AT transition mutations have been found in DNA extracted from N-methyl-N-nitrosourea-induced rat mammary tumours, a number of other alkylating agent induced tumours and in transformed mammalian cells (8,11).

The importance of other DNA lesions in the carcinogenic effects of alkylating agents remains to be established. However, O^4-alkylthymine is also a mutagenic lesion (12) and from chronic administration experiments in which this lesion accumulates due to a repair deficiency it has been suggested that in some cases O^4-ethylthymine may be the principal carcinogenic lesion (8,9,13,14).

SIGNIFICANCE OF ALKYLTRANSFERASE IN CHEMOTHERAPY

Several alkylating agents are currently used for the therapy of human neoplasms, either alone or in combination with other drugs. Many of them are bifunctional alkylating agents and probably their mechanism of cytotoxicity is related to their ability to induce DNA-interstrand cross links (DNA-ISC)(15,16). In the case of nitrogen mustards or cis-dichlorodiammineplatinum DNA-ISC are located between two N7 guanine located in a GC sequence. In contrast, for chloroethylnitrosoureas the DNA-ISC are located between a guanine and a cytosine (i.e. 1-(N3deoxycytidyl), 2-(N1-deoxyguanosyl)ethane). The formation of chloroethylnitrosoures induced DNA-ISC is presumably mediated by several reactions. Ludlum's group (17) proposed that the chloroethylcarbonium ion attacks the O^6 position of deoxyguanosine forming O^6 chloroethyldeoxyguanosine which rearranges and subsequently crosslinks with the opposite deoxycytidine. The first electrophilic attack on O^6 guanine occurs very quickly, whereas the formation of DNA-ISC takes several hours (16). Therefore an efficent repair of O^6 guanine alkylation prevents the formation of DNA-ISC. It has in fact been shown that in cells with high levels of ATase, chloroethylnitrosoureas form much less DNA-ISC than in cells with low ATase content (for a review see 18). Since the formation of DNA-ISC appears crucial for the cytotoxicity of chloroethylnitrosoureas (16,18) it seems conceivable that high cellular levels of ATase confers resistance to chloroethylnitrosoureas. The inverse relationship between ATase content and sensitivity has been found for chloroethylnitrosoureas but not for other crosslinking agents such as nitrogen mustard or cisplatinum. However, this can be explained by the fact that for these drugs the O^6 guanine alkylation is not involved in the mechanism of DNA-ISC formation.
For methyltriazenes, which are the active metabolites of dimethyltriazenes (e.g. DTIC), which are currently used clinically, there is increasing evidence that ATase content is important for the cellular resistance (19), thus suggesting that the O^6 guanine alkylation by these drugs is important for their cytotoxicity and antitumor activity. As will be discussed later, on the basis of the available knowledge it seems conceivable that patients affected by tumours with very high levels of ATase should not receive chloroethylnitrosoureas or methyltriazenes because they will not benefit from treatment and only be exposed to adverse reactions. However proper studies, designed to assess the relationship between tumour levels of ATase and clinical response to chloroethylating and methylating agents are still

missing. We are still ignorant about the values of ATase to be considered "high" or "low" for each human tumour. Without these studies it will be obviously impossible to recommend patients admission or exclusion from treatment according to ATase tumour levels.

Considering that ATase is a suicide enzyme it is theroretically possible to increase the antitumour activity of chloroethylnitrosoureas by depleting the levels of repair enzyme in the tumour. O^6-methylguanine (O^6-MeG) or methylating agents can in fact cause a depletion of ATase in tissues (9, 10, 18). This may have obvious implication for combination chemotherapies which could be based on this principle.

A recent observation indicates that in addition to methyltriazenes or to procarbazine cisplatinum can also inactivate ATase (20). On the other hand 5-azacytidine appears to increase ATase levels in murine sarcoma virus transformed NIH 3T3 cells (21), suggesting the possibility that this drug could antagonize the cytotoxic effect of chloroethylnitrosoureas or methyltriazenes. A better knowledge on the effects of the available anticancer agents on ATase levels is warranted in order to combine drugs in a rational way.

Since the goal to be persued is the increase in the therapeutic index of antitumour agents causing an initial critical lesions at O^6-guanine (i.e. chloroethylating or methylating agents) research should be addressed either at increasing the levels of ATase in normal tissues (e.g. in bone marrow) which are the target of drug toxicity or at decreasing the levels of the enzyme in tumour cells. Alternatively locoregional therapies with agents which can effectively cause depletion of ATase in tumour cells (but hopefully not in bone marrow) could be combined with systemic chloroethylnitrosoureas treatment. In this respect CNS tumours should be potential interesting candidates for this approach, since local intraarterial treatment is feasable and nitrosoureas are the drugs of choice currently in use because of their ability to cross the blood brain barrier.

METHODS OF ASSAY OF ALKYLTRANSFERASE

Both mammalian and bacterial ATases transfer alkyl groups from the O^6-position of guanine in alkylated DNA (or from the free base, but at a vastly lower rate) to a cysteine residue located at the active site of the protein: the reaction is stoicheiometric and the protein is irreversibly inactivated by the transfer (see Margison et al, this volume). This mechanism has been exploited in the design of several different radioactivity-based assays for the enzyme. These involve either measurement of the formation of S-methylcysteine in the ATase protein itself (i.e. methyl group transfer methods) or the analysis of methylated substrate DNA or synthetic oligonucleotides before and after exposure to the enzyme.

The small amounts of ATase protein, even in cells or tissues that express relatively high levels (0.001% of total protein) dictates the use of high specific radioactivity substrates. Tritium is most frequently used but more recently [^{32}P] or [^{35}S]-based assays have been developed (see below).

[^{3}H]-methyl transfer methods

Assays involving the measurement of O^6-MeG in in vitro methylated substrate DNA have included HPLC separation of normal and (radiolabelled) methylated purine bases following acid hydrolysis (e.g. ref 22). In these methods, the decrease in the O^6-MeG peak height is directly proportional to the amount of ATase added. O^6-MeG in acid, or O^6MedG in enzymic hydrolysates of such DNA have also been quantitiated following precipitation using anti-O^6-MedG antibodies (e.g. ref 23).

The most common methods of measuring ATase activity in cell or tissue extracts avoid the use of chromatography by measuring [^{3}H]-methyl group transfer to ATase protein. Although this has been achieved by digesting the protein rather than the substrate DNA to acid solubility using proteases, the most frequently used method is to hydrolyse the DNA to acid solubility and to measure radioactivity in residual protein. This can most conveniently be achieved by heating in the appropriate strength of perchloric or trichloroacetic acid for 30-45 mins at 75°C since the [^{3}H]-methylated ATase is resistant to such treatment (see ref 25,26,27). The convenience of this assay is further improved because the incubation step, acid hydrolysis, centrifugation and scintillation counting can all be achieved without involving any transfers. Protein precipitates have also been collected by filtration, although transfer efficiency and possibly quenching of radioactivity could present problems.

[^{3}H]-methylated substrate DNA is generally produced by addition of [^{3}H]-MNU (1-30 Ci/mmole) to DNA in solution followed by clean-up. As proteins are also methylated by MNU, the DNA should be as free of protein as possible. Calf thymus or <u>Micrococcus luteus</u> DNA has been used, the latter because of its very high GC content which increases its reactivity with MNU. In some cases methylated DNA's have been heated at submelting temperatures to release labile methylated bases but this would appear to be unneccessary if the protein transfer method is used.

The details of these procedures are available in the corresponding publications (22-26).

Some Technical Observations

ATase activity can be assayed in cells and tissues using total or nuclear extracts prepared by homogenisation and/or sonication and/or salt extraction. Few comparative studies have been carried out to establish which of these is most valid or reproducible. It might be argued that, provided the method is used in a consistent fashion, conclusive information should be obtained. On the other hand, comparison of results obtained in different laboratories would be more convincing if the same method was adopted.

In assays using rat liver nuclear extracts or L1210 whole cell homogenates, higher amounts of extract resulted in lower ATase activity (Fig 1). Similar results have been obtained using total mouse liver sonicates and in human tumour xenografts (unpublished results). Although the reasons for this decrease remain to be established (see below) it is likely that they are not important in as much as the ATase specific activity is (and can only be) calculated from the linear part of the protein-dependence curve. Since a linear region is almost invariably produced it is reasonable to conclude that if the above observation is the result of inhibitory factor(s) they do not affect specific activity calculations.

The possible contribution of proteases in the extract have been assessed by adding extraneous protease (XIV) to the incubation mixture. Increasing amounts of protease resulted in a drastic decrease in ATase activity (Fig 2). This indicates that high protease activity in cell or tissue extracts may, if not effectively inhibited, affect the results. If this is the explanation for the results in Fig 1 it may be advisable to use a cocktail of protease inhibitors routinely in addition to the PMSF normally added immediately after extract preparation (see also below).

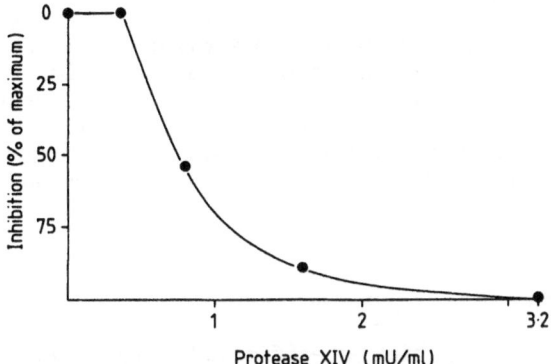

Figure 1 Alkyltransferase activity in A) crude nuclear extracts of Fischer 344 rat liver and B) crude whole homogenates of mouse L210(0) or L1210 BCNU(0) cells.

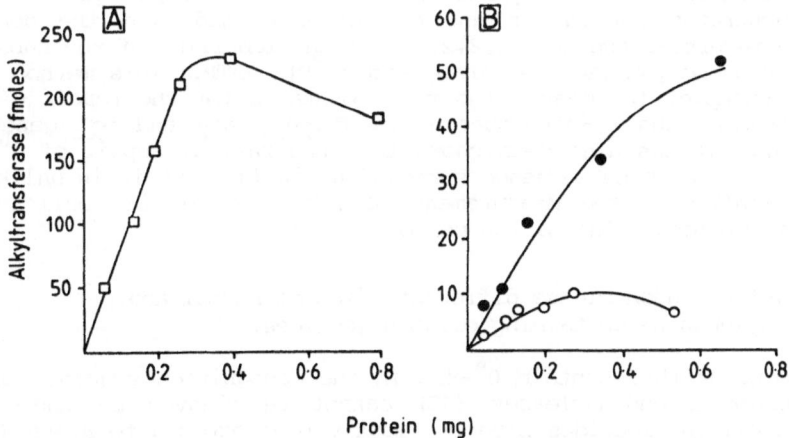

Figure 2 Reduction of alkyltranferase activity in crude nuclear extracts of Fischer 344 rat liver by simultaneous incubation with increasing amounts of protease XIV

If the comparisons of ATase activities in tissues and tumour types are to be made between different laboratories, standardisation of methods of extract preparation and assay procedures etc will be essential. To this end, standardised bacterial ATase and [^{3}H]-methylated substrate DNA can be made available to groups willing to share results with participating laboratories (contact G.P. Margison). We would recommend the preparation of supernatants from total cell or tissue sonicates, the use of the protease inhibitors PMSF and caproic acid and the assay involving perchloric acid hydrolysis (26, see later). It should be emphasised that whilst this assay is not as sensitive as those described below it is adequate for a number of tumour biopsies and is extremely easy to perform and generally highly reproducible.

[^{32}P] based methods

The availability of very high specific activity [^{32}P] and [^{35}S] labelled deoxynucleoside triphosphates has encouraged the development of more sensitive ATase assays based on end labelled DNA fragments or oligonucleotides containing 0^6-alkylguanine. A number of different methods have been devised with, theoretically, almost no limit to their sensitivity since runs of [^{32}P] labelled residues can be produced in DNA or oligonucleotides.

a) Antibody precipitation of alkylated plasmid DNA

Restriction endonuclease digested, end-filled (using [^{32}P] deoxynucleoside triphosphates,) N-ethyl-N-nitrosourea ethylated plasmid DNA can be precipitated using anti-0^6-ethyldeoxyguanosine antibodies. Preincubation with extracts containing ATase renders them non-precipitable to an extent that is proportional to the amount of enzyme added (26).

b) HPLC separation of normal and alkylated oligonucleotides

Self-complementary dodecadeoxyribonucleotides containing alkylated bases can act as substrates for ATases and can be separated from the dealkylated product by reverse-phase HPLC (28,29). 5' end-labelling of the substrate is acheived using polynucleotide kinase and [^{32}P] -ATP. This method has been used (for example) to measure the rate constants for the repair of 0^6-MeG, 0^4-methylthymine and 0^6-ethylguanine by E.coli, ada and ogt gene encoded ATases (30). It has also been shown that the rate of repair of 0^6-MeG can be affected by the base sequence surrounding it (31) but it is unlikely that this will influence the measurement of ATase levels in cell or tissue extracts if prolonged incubation periods are used.

c) Restriction endonuclease differentiation of normal and
    0^6-methylguanine containing oligonucleotides.

Oligonucleotides that contain 0^6-MeG in the recognition sequence of one or more restriction endonucleases (RE) cannot be cleaved by these enzymes unless the oligonucleotides have previously been exposed to ATase (30). 5' labelling is achieved as above. Cleaved and intact oligonucleotides can be separated by PAGE, visualised by autoradiography and quantitiated by densitometry. Extracts of ATase positive and negative cell lines have been used to confirm the validity of this assay (32).

d) Covalent binding assay

Short oligonucleotides reacted with chloroethylating agents such as BiCNU covalently bind ATase via 0^6-chloroethylguanine residues that have undergone cyclisation to N1-0^6-ethanoguanine (33,34). End-labelling is acheived as above although in this case [^{35}S] has sometimes been used because of its longer half-life. The free and ATase-bound oligonucleotides can be separated by PAGE, visualised by autoradiography and quantitated by scanning densitometry. One disadvantage of this method is that the cyclic guanine derivative is unstable and on formation of interstrand crosslinks is no longer a substrate for the ATase (33,34). A variation on this assay involves precipitation of labelled chloroethylated DNA using potassium-SDS only if the ATase protein is covalently bound to it.

ALKYLTRANSFERASE LEVELS IN HUMAN TISSUE SAMPLES

The [^{3}H]-methyl group transfer method involving hydrolysis of residual substrate DNA to acid soluble nucleotides using perchloric acid has been used to measure the amounts of ATase in human tumours and normal tissues from the same patient.

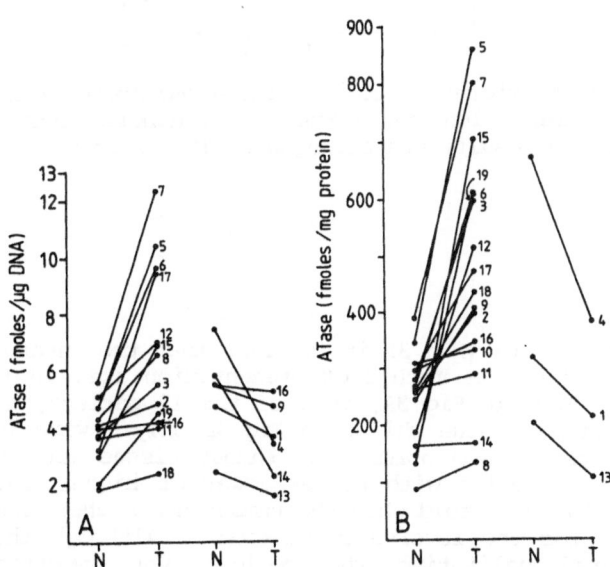

Figure 3 Alkyltransferase activity in extracts of human colon samples. A, fmoles/µg DNA; B, fmoles/µg protein. N denotes normal and T tumour tissue. For Clarity, results are arranged as T higher than N or N higher than T. Numbers refer to patient number.

The results of two sets of experiments are presented. In Fig 3A and 3B the extracts were prepared from colon samples by polytron homogenisation followed by sonication. Aliquots of the sonicate were taken for estimation of DNA content using Hoechst 33258 (bisbenzamide). Following the addition of the protease inhibitors PMSF and caproic acid the samples were centrifuged and the supernatant was used to estimate protein and ATase activity. In Fig 4, colon or stomach samples were homogenised using an ultraturrax and, following centrifugation, the supernatant was used for both protein and ATase estimation.

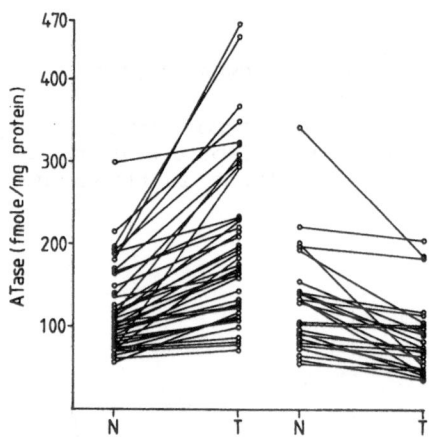

Figure 4   Alkyltransferase activity (fmoles/mg protein) in extracts of human colon and stomach.   N denotes normal and T tumour tissue and results are arranged as T>N or N>T.

In the first set of data (Fig 3) it is clear that the interpretation of the results is affected by the method of presentation i.e. in terms of DNA or protein content. Thus in Fig 3A, in 6 of the 19 patient, tumour ATase was lower than the normal tissue whilst in Fig 3B only 3 were lower. Also the maximum differences between normal and tumour tissue were lower in Fig 3A (3-fold for tumour samples with either lower or higher ATase than normal tissue) than in Fig 3B (5-fold when the tumour had higher than normal tissue ATase and 2-fold for the reverse situation). Although these differences might be considered qualitative, they could be very important since it may be that relatively small differences in ATase acitivity will be very significant in terms of tumor response (see below). Since DNA-based results will allow calculation of ATase activity per diploid cell and since some tumours, especially those of the breast are very acellular, presentation of data in terms of tissue DNA content is to be encouraged.

The second set of data (Fig 4) are for 68 colon and stomach samples and comparison with Fig 3 indicates generally qualitatively similar data in terms of the extents of difference between normal and tumour tissue.

However, the proportion of tumour ATase values which are less than normal values (38%) is much higher in this larger sample number. Also the mean ATase levels found in the tumour and in the normal tissue are about half those in Fig 3. These quantitative differences may well represent sample spread but this has yet to be established.

These results have simply demonstrated the feasibility of measuring ATase levels in human tissues using this particular assay procedure. The significance of the ATase levels with respect to the response of tumour or normal tissue to the biological effects of alkylating agents remains to be established. However, it is interesting to note that in chloroethylating agent sensitive transplantable mouse sarcomas, ATase levels were considerably lower (15 fm/mg) than in any of the human samples considered here and furthermore the levels in chloroethylating agent resistant mouse tumours (up to 70 fm/mg) were only 3-5 times higher than in the sensitive tumours (35): these latter values were only approaching the lowest levels found in the human samples. If ATase levels are the major factor in susceptibility to choroethylating agents then it might be predicted that vastly higher levels of DNA damage will need to be produced in human than in mouse tumours in order to be therapeutic.

CONCLUSIONS

Although extensive correlative evidence indicates that ATase may be an important factor in the protection of animal tissues against the carcinogenic effect of alkylating agents it will, of course, be difficult to arrive at any conclusions for human cancer induction. 0^6-Alkylguanine has been shown to be present in human DNA and this will eventually be assessed in epidemiological studies as another possible risk factor in carcinogenesis. Proof of cancer causation by 0^6-alkylguanine or any other DNA lesion may, because of the complexity of the problem be difficult to obtain.

In terms of chemotherapy by the types of agent that produce 0^6-alkylguanine in DNA it is now clearly time to assess ATase levels in biopsy material so that these values can, wherever possible be related to the response of the tumour to treatment. It should be emphasised that, extrapolating from the mouse data, relatively small differences in ATase levels, similar in magnitude to those seen in the human samples will be clinically very relevant. Equally important in terms of unwanted haematological side-effects would be to assess enzyme levels in bone marrow aspirates. It may be that in some patients, reduced myelosuppression may be a consequence of protection by relatively high ATase levels in critical target cells. Although other explanations are possible, circumstantial evidence suggests that ATase may be a contributory factor.

For the human tumours examined so far, the amount of material that can be available following biopsy is, even at present, sufficient to measure ATase levels (see above). If such large (0.25-4g) samples can be obtained, collaborating centres in Bern, London, Newcastle, Manchester and Milan will perform or demonstrate the [^{3}H]-methyl transfer assay to groups willing to share results (contact Cerny, D'Incalci or Margison at address given).

It is expected that the highly sensitive assays described briefly here will require much smaller samples i.e. needle biopsies, in order to generate reliable data. The question of tissue heterogeneity within the tumour and between tumour nodules could be addressed in this way with cytological confirmation that tumour material was being examined. The ease of use and reproducibility of the sensitive assays is being explored and it may be that variations on one or more of the methods outlined will eventually provide the most convenient method. The results of such assays should be expressed in terms of DNA content for the reasons outlined above and because of the much greater sensitivity of the DNA estimation method.

Meanwhile, it is hoped that a large number of cancer centres which use methylating and/or chloroethylating agents will take and store pretreatment tumour samples or needle biopsys for determination of their ATase levels by appropriate methods when they enter routine use. Only in this way can the possible correllation between ATase expression and tumour response be assessed and this will establish the usefulness of ATase levels as a prognostic factor. In addition a better knowledge of the levels and the turnover rate of ATase in human normal and neoplastic tissues as well as of the kinetic of depletion of ATase by inactivating agents in the same tissues may provide the necessary biochemical basis to attempt more rational combinations.

ACKNOWLEDGEMENTS

We gratefully acknowledge support from the Cancer Research Campaign (UK)

REFERENCES

1. Lawley, P.D. (1972) The action of alkylating agents on nucleic acids: N-methyl-N-nitroso compounds as methylating agents. In W N Kakahara et al (eds) Topics in chemical carcinogenesis, Univ of Tokyo Press, Tokyo, pp 237-256

2. Singer, B. (1975) The chemical effects of nucleic acid alkylation and their relation to mutagenesis and carcinogenesis. Prog Nucl Acids Res Molec Biol 15, 219

3. Pegg, A.E. (1977) Metabolism of alkylated nucleosides: possible role in carcinogenesis by nitroso compounds and alkylating agents. Adv Canc Res 25, 195-270

4. Roberts, J.J. (1978) The repair of DNA modified by cytotoxic, mutagenic and carcinogenic chemicals. Adv Radiat Biol 7, 211-436

5. Singer, B. (1979) N-Nitrosoalkylating agents: Formation and persistence of alkyl derivatives in mammalian nucleic acids as contributing factors in carcinogenesis. J. Natl. Cancer Inst. 62, 1329-1339

6. Margison, G.P. and O'Connor, P.J. (1979) Nucleic acid modification by N-nitroso compounds. In P.L. Grover (ed) Chemical Carcinogenesis and DNA. CRC Press, Boca Raton, pp 111-159

7. Singer, B. and Kusmierek, J.T. (1982) Chemical mutagenesis. Ann Rev Biochem 52, 655-693

8. Saffhill, R., Margison, G.P. and O'Connor, P.J. (1985) Mechanisms of carcinogenesis induced by alkylating agents. Biochem Biophys Acta 823, 111-145

9. Margison, G.P. and O'Connor, P.J. (1989) Biological consequences of reactions with DNA: Role of specific lesions. In P.L. Grover and D.H. Phillips (Eds) Chemical Carcinogenesis and Mutagensis Handbook of Experimental Pharmacology Vol 94/1. Springer, Heidelberg, In Press.

10. Day, R.S., Babich, M.A., Yarosh, D.B. and Scudiero, D.A. (1987) The role of O^6-methylguanine in human cell killing sister chromatid exchange induction and mutagenesis: a review. J. Cell Sci. Suppl. 6, 333-353

11. Sukumar, S., Notario, V., Martin-Zanca, D. and Barbacid, M. (1983) Induction of mammary carcinomas in rats by nitroso-methylurea involves malignant activation of H-ras-1 locus by single point mutation. Nature (London), 306, 658-662

12. Abbott, P.J. and Saffhill, R. (1977) DNA synthesis with methylated poly(dA-dT) templates: Possible role of O^4-methylthymine as a promutagenic base. Nucleic Acids Res 4, 761-769

13. Swenberg, J.A., Dyroff, M.C., Bedell, M.A., Popp, J.A., Huh, N. Kirstein, U. and Rajewsky, M.F. (1984) O^4-Ethyldeoxythymidine but not O^6-ethyldeoxyguanosine accumulates in hepatocyte DNA of rats exposed continuously to diethylnitrosamine. Proc. Natl. Acad. Sci. (U.S.A.). 81, 1692-1695

14. Scherer, E.A.P., Timmer and P. Emmelot (1980) Formation by diethylnitrosamine and persistence of O^4-ethylthymidine in rat liver DNA in vivo. Cancer Lett. 10, 1-6

15. Zwelling, L.A, Michaels, S., Schwartz, H., Dobson, P.P. and Kohn, K.W. (1981) DNA crosslinking as an indicator of sensitivity and resistance of mouse L1210 leukemia cells to cis-diamminedichloroplatinum (11) and L-phenylalanine mustard. Cancer Res. 41, 640-649

16. Erickson, L.C., Zwelling, L.A., Ducore, J.M., Sharkey, N.A. and Kohn, K.W. (1981) Differential cytotoxicity and DNA crosslinking in normal and transformed human fibroblasts treated with cis-diamminedichloroplatinum (11). Cancer Res. 41, 2791-2794

17. Tong, W.P., Kirk, M.C. and Ludlum, D.B. (1982) Formation of the cross-link 1-(N3-deoxycytidyl), 2-N1-deoxyguanosyl)-ethane in DNA treated with N, N-bis(2-chloroethyl)-N-nitrosourea. Cancer Res. 42, 3102-3105

18. D'Incalci, M., Citti, L., Taverna. P. and Catapano, V.C. 81988) Importance of the DNA repair enzyme O^6-alkylguanine alkyltransferase (AT) in cancer chemotherapy. Cancer Treat Review. 15, 279-292

19. Catapano, V.C., Broggini, M., Erba, E., Ponti, M., Mariani, L., Citti, L. and D'Incalci, M. (1987) In vitro and in vivo methazolastone-induced DNA damage and repair in L1210 leukemia sensitive and resistant to chloroethylnitrosoureas. Cancer Res. 47, 4884-4889

20. Wang, L. and Setlow, R.B. (1989) Inactivation of O^6-alkylguanine- DNA alkyltransferase in Hela cells by cisplatin. Carcinogenesis 10, 1681-1684

21. Mitani, H., Yagi, T., Leiler, C.Y.M. and Takebe, H. (1989) 5-Azacytidine-induced recovery of O^6-alkylguanine-DNA alkyltransferase activity in mouse Ha821 cells. Carcinogenesis 10, 1879-1882

22. Cooper, D.P., O'Connor, P.J. and Margison, G.P. (1982) Effect of acute doses of 2-AAF on the capacity of rat liver to repair methylated purines in DNA in vivo and in vitro. Cancer Res. 42, 4203-4209

23. Pegg, A.E., Wiest, L., Foote, R.S., Mitra, S. and Perry, W. (1983) Purification and properties of O^6-methylguanine-DNA transmethylase from rat liver. J. Biol. Chem. 258, 2327-2333.

24. Waldstein, E., Cao, E.H. and Setlow, R.B. (1983) Direct assay for O^6-methylguanine acceptor protein in cell extracts. Anal. Biochem. 126, 268-272

25. Myrnes, B., Nordstrand, K., Giercksky, K.E., Sjunneskog, C. and Krokan, H. (1984) A simplified assay for O^6-methylguanine-DNA methyltransferase activity and its application to human neoplastic and non-neoplastic tissues. Carcinogenesis 5, 1061-1064

26. Margison, G.P., Cooper, D.P. and Brennand, J. (1985) Cloning of the E.coli 0^6-methylguanine and methylphosphotriester methyltransferasegene using a functional DNA repair assay. Nucleic Acids Res 13, 1939-1952

27. Nrehls, P. and Rajewsky, M.F. (1989) Monoclonal antibody-based immunoassay for the determination of cellular enzymatic activity for repair of specific carcinogen-DNA adducts (0^6-alkylguanine). Carcinogensis. In Press.

28. Graves, R.J., Li, B.F.L. and Swann, P.F. (1989) Repair of 0^6-methylguanine, 0^6-ethylguanine, 0^6-isopropylguanine and 0^4-methylthynine in synthetic oligodeoxynucleotides by E.coli ada gene 0^6-alkylguanine-DNA-alkyltransferase. Carcinogenesis 10, 661-666

29. Dolan, M.E., Scicchitano, D. and Pegg, A.E. (1988) Use of oligodeoxynucleotides containing 0^6-alkylguanine for the assay of 0^6-alkylguanine-DNA-alkyltransferase activity. Cancer Res. 48, 1184-1188

30. Wilkinson, M.C., Potter, P.M., Cawkwell, L.,Georgiadis, P., Patel, D. Swann, P.F. and Margison, G.P. (1989) Purification of the E.coli ogt gene product to homogeneity and its rate of action on 0^6-methylguanine, 0^6-ethylguanine and 0^4-methylthymine in dodecadeoxyribonucleotides. Nucleic Acids Res. 17, 8475-8484

31. Topal, M.D., Eadie, J.S., and Conrad, M., 0^6-methylguanine mutation and repair is nonuniform. Selection for DNA most interactive with 0^6-methylguanine, J. Biol. Chem. 261, 9879, 1986

32. Wu, R.S., Hurst-Calderone, S. and Kohn, K.W. (1987) Measurement of 0^6-alkylguanine-DNA-alkyltransferase activity in human cells and tumour tissues by restriction endonuclease inhibition. Cancer Res. 47, 6229-6235

33. Brent, T.P., Smith, D.G. and Remack, J.S. (1987) Evidence that 0^6-alkylguanine-DNA alkyltransferase becomes covalently bound to DNA containing 1,3-bis(2-chloroethyl)-1-nitrosourea-induced precursors of interstrand cross-links Biochem. Biophys. Res. Comm. 142, 341-352

34. Brent, T.P. and Remack, J.S. (1988) Formation of covalent complexes between human 0^6-alkylguanine-DNA alkyltransferase and BCNU-treated defined length synthetic oligonucleotides. Nucleic Acids Res. 16, 6779-6788

35. Kooistra, K., Donaldson, J., Workman, P. and Margison .G.P. (1988) Br. J. Cancer

206

E. Bonmassar[1], A. Gescher[2], D.R. Newell[3], M.F.G. Stevens[4]

[1] Department of Experimental Medicine and Bio-chemical Sciences, II University of Rome, Italy [2] Cancer Research Campaign, Pharmaceutical Sciences Institute, Aston University, Birmingham, U.K.; [3] Institute of Cancer Research, Sutton Surrey, England and [4] Pharmaceutical Sciences, Aston University, Birmingham, U.K.

A. Varnavas and colleagues (16) have synthesized 1,3-di(4-carboxyphenyl)triazene (I), which is a diaryltria-zene analogue of the antimetastatic agent DM-COOK (II).
Whereas DM-COOK can be converted to MM-COOK (III), the (presumed) active methylating metabolite, this route of activation is not available to the diaryltriazene. On the other hand, the diaryltriazene, like DM-COOK, can undergo breakdown to the diazonium ion (IV). Compound (I) has been tested against murine TLX5 lymphoma and Lewis lung carcinoma and shown to have no antitumoral activity. This result confirms that the diazonium species (IV) is not involved in the biological activity of DM-COOK.

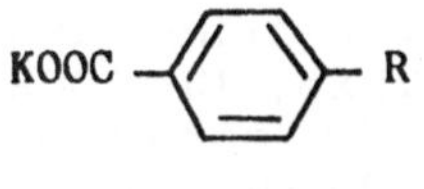

$$KOOC - \langle \bigcirc \rangle - R$$

(I)   R = $N{=}N{-}NHC_6H_4CO_2K(p)$
(II) R = $N{=}N{-}NMe_2$
(III) R = $N{=}N{-}NHMe$
(IV) R = N_2^+

A. Varnavas and colleagues (17) also synthesized a series of analogues of DM-COOK (II) where the carboxylic acid residue was separated from the aryl ring by a one-, two- or three- chain carbon fragments.

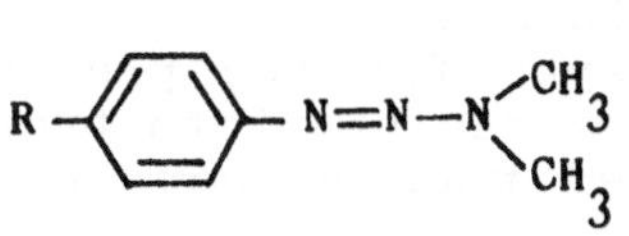

$$R - \langle \bigcirc \rangle - N{=}N{-}N\langle {}^{CH_3}_{CH_3}$$

(V)   R = CH_2CO_2K
(VI) R = $CH{=}CHCO_2K$
(VII) R = $(CH_2)_3CO_2K$

All four compounds had pKa values in the 4.5 - 4.7 range and (as neutral molecules) varied in a *log P* from 1.38 for compound (II) to 2.66 for the more lipophilic compound (VI).

DM-COOK was the most active compound in inhibiting lung metastasis in mice bearing the Lewis lung carcinoma, and the carbon-extended analogues were more toxic, but significantly less antimetastatic.

A poster from the Semmelweis Medical University in Budapest (11) focussed on the antimetastatic effect of the triazenes DTIC and DM-COOK. Lewis lung carcinoma with differing metastatic potential were chosen as experimental models. The inhibitory effect of either drug on the number of metastasis was dependent on both the metastasis-forming capacity of the tumor and the presence or absence of the primary tumor.

When given at different dose schedules, only continuous treatment over 14 days was antimetastatic. The results show that triazenes are able to influence different steps of metastasis progression, and their effect is dependent on dose scheduling and tumor type.

A poster by Sava et al. (13) was concerned with the metabolic N-demethylation *in vitro* as measured by HCHO release of a series of halogenated N,N-dimethylphenyltriazenes. All of the compounds underwent similar levels of overall demethylation (50-65%) but the initial rate was faster for those agents with only a single substituent (Cl,Br,CF₃) *para* to the triazene function relative to aryl poly-halogenated compounds (2,4,6-Br or Cl). The slower release of the HCHO from 2,4,6 Br or Cl aryltriazenes may reflect the greater stability of the hydroxymethyl intermediates. It was suggested that the greater stability of the hydroxymethyl intermediate may allow conjugation at this position and hence elimination. This could explain the reduced carcinogenity of the poly-halogenated triazenes.

The study by Sava et al. (14) addressed the biological properties of MM-COOK, the N-mono-methyl metabolite of DM-COOK. Like DM-COOK this compound could reduce the number and size of metastases from the Lewis lung carcinoma. In addition, it has activity against the primary Lewis lung carcinoma and can reduce the number and size of metastatic deposits from MCa mammary carcinoma. Against established metastatic tumors, MM-COOK is active on the Lewis lung carcinoma but not MCa. Against both i.p. and i.c. TLX5, MM-COOK and DM-COOK have similar activity but MM-COOK is more potent. The data on MM-COOK are consistent with it being an active metabolite of DM-COOK, and as such it may be more potent.

A poster (8) by Lunn et al. addressed the question whether the resistance observed in human melanomas grown in nude mice towards the water-soluble triazene CB10-277 was due to the ability of the tumour to repair lesions caused by alkylation at the guanine O⁶-position in DNA.

Extracts of the tumours were assayed for the presence of the repair enzyme. Tumours which were resistant to treatment with the triazene displayed indeed much higher enzyme levels than the sensitive tumours.

A similar problem was investigated by Taverna et al. (15). The compound under investigation was temozolomide to which L1210 mouse leukemia cells were exposed, which were derived from lines with or without acquired resistance towards temozolomide and other alkylating agents. Temolozomide caused the formation of DNA alkali-labile sites in both cell lines, which were probably caused by N^7-guanine alkylation. However, the extent of methylation of the O^6 position at guanine caused by temolozomide was higher in the sensitive than in the resistant line. Addition to the cell suspension of O^6 methylguanine, which causes depletion of O^6-alkylguanine DNA alkyltransferase, reversed the resistance of the cells to temolozomide. Separate studies showed that temolozomide depletes O^6-alkylguanine DNA alkyltransferase and can enhance the cytotoxicity of certain nitrosoureas.

Two posters addressed clinical studies with DTIC

The study by Kolaric et al. (7) was concerned with the alternating combinations of BCNU / DTIC / Vincristine and Cisplatin / Vincristine / Bleomycin for the treatment of metastatic melanoma. In this phase II study the overall response rate (29%, 5/17) was no better than would be predicted for DTIC alone. In contrast the toxicity, notably alopecia, was more than might be expected for DTIC. It is concluded by the authors that the alternating schedule was no better than either arm alone. These results suggest strongly that it is the underlying biology of the tumour which determines response rates rather than the drugs used.

In a further phase II study (Milani et al., 20) DTIC and Interferon (alpha-2b, INF) were combined. In 22 patients there were 11 partial responses, a result which is encouraging. Toxicities of the combination (haematologic, nausea, astenia, myalgia) were felt to be acceptable and half the patients showed a partial response. This interesting result requires confirmation and extension to cover the questions of dose, dose intensity and the exact role of the IL2; phase III studies of single agent DTIC, IL2, the combination would be one way to proceed.

Six poster presentations were related to the immuno-pharmacological properties of triazenes

Two of them by Cartei et al. (2, 3) were concerned with the immunodepressant effects of DTIC in man. High doses of the drug given for 3-4 cycles to cancer-bearing patients seem to produce a certain degree of impairment of T lymphocyte subsets. However 2 cycles of high dose DTIC or chronic administration of limited doses of the drug has little or no influence on white cells and lymphocyte subsets, possibly involved in cell-mediated immunosurveillance.

The other 4 posters described the phenomenon of "chemical xenogenization" (CX, i.e. induction of novel antigenic specificities on mammalian cells by in vivo or in vitro treatment with drugs) produced by triazenes in mouse or human cancer cells.

Fuschiotti et al. (6) performed a comparative analysis of CX in mouse models induced by triazenes and other chemically-unrelated compounds with similar mechanism of action (i.e. somatic mutations produced by methylating agents such as nitrosoguanidines, via generation of O^6-guanine methyl adducts in DNA) or with different mechanism, such as DNA hypomethylation (i.e. epigenetic phenomenon produced by 5 azacytidine). The results of their study point out that triazene-mediated CX is by far stronger then that obtainable with other CX-inducing agents.

The work of D'Atri et al. (4) provides strong support to the hypothesis that CX of human lung cancer cells can be induced in vitro following exposure to triazene compounds. Cytotoxic T lymphocytes (CTL) were generated in vitro against the triazene-treated cells using allogeneic mononuclear cells of healthy donors as responder lymphocytes. CTL were cloned and a set of selected clones was found to be preferentially cytolytic for triazene-treated tumor cells, being almost inactive against the parental line. These results are consistent with the hypothesis that treated cells express membrane antigen(s) not present in the line of origin, and recognized by limited CTL clones within the context of antigens of allogeneic histocompatibility antigens of the major istocompatibility complex (i.e. allorecognition of cell membrane antigens generated by CX).

Bianchi et al. (1) point out that highly immunogenic triazene-treated cells are capable of cross-protecting mice from challenge with non-treated parental cells. In particular lymphocytes of L3T4 phenotype (i.e. T cells associated with DTH responses) collected from mice sensitized against xenogenized leukemia cells, are capable of inhibiting the growth of parental tumor in syngeneic hosts. A possible role of activated macrophages is suggested, thus providing rational bases for adoptive immunotherapy in the clinic.

Studies in human models performed by Vernole et al. (18) point out that an in vitro active monomethyl-aryl-triazene, highly efficient in inducing CX, appears to be clastogenic in human lymphocytes in G2 phase from normal donors stimulated in vitro with PHA. If lymphocytes are collected from ataxia telangectasia patients, triazene-mediated DNA damage appears to be more severe. Since interaction of triazenes with DNA topoisomerase II seems to be ruled out, the production of oxidants remains a reasonable candidate to explain the effects of the drug on cells in G2 phase. One of the most important problems related to CX concerns the possibility of inducing this phenomenon in human cancer cells for clinical application.

POSTER TITLES

1. Mechanisms of the cross-protective immunity induced by xenogenized against parental cells.
   R. Bianchi, M.C. Fioretti, L. Romani, M. Allegrucci and P. Puccetti.
   Dept. Exp. Med. and Bioch. Sci., Pharmacology Section, University of 06100 Perugia, Italy.
2. Effects of Dacarbazine (DTIC) on immunology in the human: I: High intermittent dose (HID).
   G. Cartei[1], T. Giraldi[2] and R. Carella[1].
   [1] Divisione di Oncologia Medica, Ospedale Regionale, Udine and [2] Istituto di Farmacologia, Università di Trieste.

3. Effect of Dacarbazine (DTIC) on immunology in the human: II: Low and intermediate chronic dose schedule (LDC,ICD).

G. Cartei[1], T. Giraldi[2] and R. Carella[1].
[1] Divisione di Oncologia Medica, Ospedale Regionale, Udine and [2] Istituto di Farmacologia, Università di Trieste.

4. Allorecognition of antigenic modifications induced by 4 (3-methyl-1-triazeno) benzoic acid (MTBA) in a human tumor line.

S. D'Atri[1], M. Tricarico[2] and E. Bonmassar[1].
[1] Dep. of Experimental Medicine and Biochemical Sciences. II University of Rome, Rome, Italy and [2] Institute of Experimental Medicine, CNR, Rome Italy.

6. Comparative analysis of xenogenization patterns by chemicals belonging to different classes.

P. Fuschiotti, U. Grohmann, L. Romani, P. Puccetti and M. C. Fioretti.
Dept. Exp. Med. and Bioch. Sci., Pharmacology Section, University of 06100 Perugia, Italy.

7. Combination of BCNU DTIC VCR alternating with cis DDP Vinblastin and Bleomycin as primary chemotherapy in metastatic melanoma– a phase II trial.

K. Kolaric, R. Tomek, Z. Mrsic and D. Zupanc.
Central Institute for Tumors and Allied Diseases, Zagreb, Yugoslavia.

8. Correlation between chemosensitivity to CB10-277 and O[6]-alkyl-guanine-DNA alkyltransferase levels in human melanoma xenografts.

J. M. Lunn[1], B. J. Foster[2], M. Jones[2], J. Siraky[2] and D. R. Newell[2].
[1] Cancer Research Unit, University of Newcastle upon Tyne, Newcastle upon Tyne NE2 4HH, England and[2] Clinical Pharmacology, Institute of Cancer Research, Sutton, Surrey SM2 5NG., England.

11. Antimetastatic effect of dimethyltriazenes in mice bearing Lewis lung carcinoma of different metastatic potential.

E. Raso, K. Lapis and L. Kopper.
I. Insitute of Pathology and Experimental Cancer Research, Semmelweis Medical University, Ulloi ut 26, Budapest, H-1085 Hungary.

13. Activation and oxidative (MFO) N-demethylation of ring halogenated 3,3-dimethyl-1-phenyltriazenes.

G. Sava[1], S. Pacor[1], F. Bregant[1], V. Ceschia[1] and G. F. Kolar[2].
[1] Institute of Pharmacology, Faculty of Pharmacy, University of Trieste, I-34100 Trieste, Italy and [2] Institute of Toxicology and Chemotherapy, German Cancer Research Center, D-6900 Heidelberg, F.R.G.

14. Antineoplastic action of p-(3-methyl-1-triazeno) benzoic acid potassium salt, monomethylderivative of the antimetastatic compound DM-COOK.

G. Sava[1], S. Zorzet[1], L. Perissin[1], S. Pacor[1], V. Ceschia[1], F. Bregant[1], T. Giraldi[1], A. Varnavas[2], C. Nisi[2] and L. Lassiani[2].
[1] Institute of Pharmacology and [2] Institute of Pharmaceutical Chemistry and Toxicology, Faculty of Pharmacy, University of Trieste, Italy.

15. Relation between methyltriazenes cytotoxicity and cell line ability to repair O[6] alkylguanine.

P. Taverna[1], C. V. Catapano[1], L. Citti[2], M. Bonfanti[1] and M. D'Incalci[1].
[1] Istituto "Mario Negri", I-20157 Milano and [2] CNR Istituto Mutagenesi e Differenziamento, I-56100 Pisa.

16. Antitumor activity and aqueous stability of AVIS, a hydrosoluble diaryltriazene, related to the antimetastatic agent DM-COOK.
A. Varnavas[1], L. Lassiani[1], C. Nisi[1], S. Zorzet[2], L. Perissin[2] and G. Sava[2].
[1] Institute of Medicinal Chemistry and [2] Institute of Pharmacology, Faculty of Pharmacy, University of Trieste, Italy.

17. Synthesis and antitumor activity of hydrosoluble triazenes analogues of DM-COOK.
A. Varnavas[1], L. Lassiani[1], C. Nisi[1], S. Zorzet[2], L. Perissin[2], S. Pacor[2], T. Giraldi[2] and G. Sava[2].
[1] Institute of Medicinal Chemistry and [2] Institute of Pharmacology, Faculty of Pharmacy, University of Trieste, Italy.

18. Chromosome and DNA damage induced by the 1-p-(3-methyl-triazeno) benzoic acid potassium salt in human cells.
P. Vernole[1], D. Caporossi[1], B. Tedeschi[1], F. Zunino[2], E. Bonmassar[3] and B. Nicoletti[1].
[1] Dip. Sanità Pubblica & Biologia Cellulare, II Università di Roma, Via O. Raimondo 00173 Roma, Italia, [2] Ist. Naz. Studio & Cura Tumori, Milano, [3] Dip. Med. Sper., II Università di Roma.

20. Dacarbazine and Interferon alpha-2b combination in advanced malignant melanoma: a phase II study.
S. Milani, G. Mustacchi, P. Sandri and M. Mansutti.
Centro Oncologico, Via Pietà 19, 34127 Trieste, Italy.

Note The abstracts of the posters whose titles are listed above, may be found in *Anticancer Research* 10 453-474 (1990).

CONTRIBUTORS

M. Allegrucci   Department of Experimental Medicine and
                       Biochemical Sciences, Section Pharmacol-
                       ogy, University of Perugia, 06100 Peru-
                       gia, Italy

R. Bianchi   Department of Experimental Medicine and
                       Biochemical Sciences, Section Pharmacol-
                       ogy, University of Perugia, 06100 Peru-
                       gia, Italy

G.R.P. Blackledge   Department of Medicine, Queen Elizabeth
                       Hospital, Birmingham B15 2TH, U.K.

E. Bonmassar   Department of Experimental Medicine and
                       Biochemical Sciences, II University of
                       Rome, 00173 Rome, Italy

J. Brozmanova   Cancer Research Institute, Slovak Acade-
                       my of Sciences, 81232 Bratislava, Czech-
                       oslovakia

K. Buser   Inselspital Bern, Institute für Medizi-
                       nische, Onkologie der Universität, CH-
                       3010 Bern, Switzerland

A.H. Calvert   Division of Oncology, Cancer Research
                       Unit, Medical School, Framlington Place,
                       Newcastle Upon Tyne, NE2 4HH, England

J. Carmicheal   University of Newcastle, Upon Tyne New-
                       castle, NE2 4HH, England

G. Cartei   Divisione di Oncologia Medica, Ospedale
                       S. Maria della Misericordia, 33100 Udine
                       Italy

C.V. Catapano   Istituto di Ricerche Farmacologiche "Ma-
                       rio Negri", Via Eritrea 62, 20157 Milan,
                       Italy

T. Cerny   Inselspital Bern, Institute für Medizi-
                       nische, Onkologie der Universität, CH-
                       3010 Bern, Switzerland

L. Citti   Istituto Mutagenesi e Differenziamento
                       CNR, Via Svezia 10, 56100 Pisa, Italy

T. Colombo   Istituto di Ricerche Farmacologiche "Ma-
                      rio Negri", Via Eritrea 62, 20157 Milan,
                      Italy

T.A. Connors   Toxicology Unit, Medical Research Coun-
                      cil, Woodmansterne Road, Carshalton,
                      Surrey SM5 4EF, U.K.

D.P. Cooper   Carcinogenesis Department, Paterson In-
                      stitute for Cancer Research Christie
                      Hospital, Wlimslow Road, Manchester M20
                      9BX, U.K.

L. Cernakova   Carcinogenesis Department, Paterson In-
                      stitute for Cancer Research Christie
                      Hospital, Manchester M20 9BX, U.K.

S. D'Atri   Institute of Experimental Medicine,
                      Italian National Research Council, II
                      University of Rome, 00173 Italy

J. Davies   Carcinogenesis Department, Paterson In-
                      stitute for Cancer Research Christie
                      Hospital, Wlimslow Road, Manchester M20
                      9BX, U.K.

M. D'Incalci   Istituto di Ricerche Farmacologiche "Ma-
                      rio Negri", Via Eritrea 62, 20157 Milan,
                      Italy

S. Eckhardt   National Institute of Oncology, PF. 21.
                      XII., Rath GY. U. 7/9, 1525 Budapest,
                      Hungary

P. Farina   Istituto di Ricerche Farmacologiche "Ma-
                      rio Negri", Via Eritrea 62, 20157 Milan,
                      Italy

M.C. Fioretti   Department of Experimental Medicine and
                      Biochemical Sciences, Section Pharmacol-
                      ogy, University of Perugia, 06100 Peru-
                      gia, Italy

B.J. Foster   Division of Oncology, Cancer Research
                      Unit, Medical School, Framlington Place,
                      Newcastle Upon Tyne, NE2 4HH, England

P. Fuschiotti   Department of Experimental Medicine and
                      Biochemical Sciences, Section Pharmacol-
                      ogy, University of Perugia, 06100 Peru-
                      gia, Italy

A. Gescher   Cancer Research Campaign, Experimental
                      Chemotherapy Research Group, Pharmaceu-
                      tical Sciences Institute, Aston Univer-
                      sity, Birmingham B4 7ET, U.K.

T. Giraldi   Istituto di Farmacologia, Università di
                      Trieste, 34127 Trieste, Italy

U. Grohmann	Department of Experimental Medicine and Biochemical Sciences, Section Pharmacology, University of Perugia, 06100 Perugia, Italy
L. Gumbrell	Division of Oncology, Cancer Research Unit, Medical School, Framlington Place, Newcastle Upon Tyne, NE2 4HH, England
C.N. Hall	Department of General Surgery, Wythenshawe Hospital, Manchester, U.K.
A.L. Harris	University of Newcastle, Upon Tyne Newcastle, NE2 4HH, England
L.C. Harris	Carcinogenesis Department, Paterson Institute for Cancer Research Christie Hospital, Wlimslow Road, Manchester M20 9BX, U.K.
K. Jenns	Division of Oncology, Cancer Research Unit, Medical School, Framlington Place, Newcastle Upon Tyne, NE2 4HH, England
I.R. Judson	Drug Development Section, Institute of Cancer Research, Clinical Pharmacology, Block E, 15 Cotswold Road, Belmont, Sutton Surrey SM2 5NG, England
K. Kleibl	Cancer Research Institute, Slovak Academy of Sciences, 81232 Bratislava, Czechoslovakia
C. Mannironi	Istituto di Ricerche Farmacologiche "Mario Negri", Via Eritrea 62, 20157 Milan, Italy
G.P. Margison	Carcinogenesis Department, Paterson Institute for Cancer Research Christie Hospital, Wlimslow Road, Manchester M20 9BX, U.K.
D.R. Newell	Division of Oncology, Cancer Research Unit, Medical School, Framlington Place, Newcastle Upon Tyne, NE2 4HH, England
E.S. Newlands	Department of Medical Oncology, Charing Cross Hospital, Fulham Palace Road, London W6 8RF, England
P.J. O'Connor	Carcinogenesis Department, Paterson Institute for Cancer Research Christie Hospital, Wlimslow Road, Manchester M20 9BX, U.K.
L. Perissin	Istituto di Farmacologia, Università di Trieste, 34127 Trieste, Italy
P. Puccetti	Department of Experimental Medicine and Biochemical Sciences, Section Pharmacology, University of Perugia, 06100 Perugia, Italy

215

V. Rapozzi — Istituto di Farmacologia, Università di Trieste, 34127 Trieste, Italy

S.M.S. Redmond — Inselspital Bern, Institute für Medizinische, Onkologie der Universität, CH-3010 Bern, Switzerland

L. Romani — Department of Experimental Medicine and Biochemical Sciences, Section Pharmacology, University of Perugia, 06100 Perugia, Italy

M. Skorvaga — Cancer Research Institute, Slovak Academy of Sciences, 81232 Bratislava, Czechoslovakia

J.A. Slack — Cancer Research Campaign, Experimental Chemotherapy Research Group, Pharmaceutical Sciences Institute, Aston University, Birmingham B4 7ET, U.K.

J. Slack — Department of Pharmaceutical Sciences, Pharmaceutical Sciences Institute, Aston University, Birmingham B4 7ET, U.K.

M.F.G. Stevens — Department of Pharmaceutical Sciences, Pharmaceutical Sciences Institute, Aston University, Birmingham B4 7ET, U.K.

N.S.A. Stuart — Department of Pharmaceutical Sciences, Aston University, Birmingham B15 2TH, U.K.

P. Taverna — Istituto di Ricerche Farmacologiche "Mario Negri", Via Eritrea 62, 20157 Milan, Italy

L. Tentori — Institute of Experimental Medicine, Italian National Research Council, II University of Rome, 00173 Italy

M.J. Tisdale — Cancer Research Campaign, Experimental Chemotherapy Group, Pharmaceutical Sciences Institute, Aston University, Birmingham B4 7ET, U.K.

M. Tricarico — Institute of Experimental Medicine, Italian National Research Council, II University of Rome, 00173 Italy

L.L.H. Tsang — Cancer Research Campaign, Experimental Chemotherapy Research Group, Pharmaceutical Sciences Institute, Aston University, Birmingham B4 7ET, U.K.

K. Vaughan — Department of Chemistry, Saint Mary's University, Halifax, N.S., Canada B3H 3C3

V. Vlckova   Department of Genetics, Comenius Univer-
                    sity 842,15 Bratislava, Czechoslovakia

D.E.V. Wilman   Drug Development Section, Institute of
                    Cancer Research, Cancer Research Cam-
                    paign Laboratory, 15 Cotswold Road,
                    Belmont, Sutton Surrey SM2 5NG, U.K.

S. Zorzet   Istituto di Farmacologia, Università di
                    Trieste, 34127 Trieste, Italy

Demethylation (continued)
  hexamethylmelamine, 175
  N-methylmelamines, 180-181
  pentamethylmelamine, 176
5-Diazoimidazole-4-carboxamide
  antitumor activity in ani-
      mals, 24-25
    adenocarcinoma 755, 25
    Ehrlich carcinoma, 25
    L1210 leukemia, 25
    sarcoma 180, 25
    Walker 256 carcinosarcoma,
      25
  cyclization to 2-azahypoxan-
      thine, 24
  reactions, 24
  synthesis, 24
DM-COOK
  antimetastatic activity in
      animals, 45-57
  antitumor activity in ani-
      mals, 99, 121
    L1210 leukemia, 99
  metabolism in humans
    conjugation, 125-127
    demethylation, 125-127
  metabolism in mice, 100,
      101, 121
    conjugation, 121
    demethylation, 121
  pharmacokinetics
    in humans, 125-127
    in mice, 121
  phase I responses, 127
  xenogenization, 80-81
1-p-(3,3-dimethyl-1-triazeno)
      benzoic acid potassium
      salt, see DM-COOK and
      CB10-277
5-(3,3-Dimethyl-1-triazeno)
      imidazole-4-carboxa-
      mide, see DTIC
DNA damage repair
  and O(6)-Alkylguanine-DNA-
      alkyltransferase,
      161-169
DNA hypomethylation
  imidazotetrazinones, 19-20
  mitozolomide, 19-20
  temozolomide, 19-20
    and xenogenization, 209-
      210
Drug combination(s) including
      DTIC, see DTIC
DTIC
  and antiemetic therapy, 136
  antitumor activity in
      animals, 26-28
    adenocarcinoma 755, 27
    L1210 leukemia, 26, 27

DTIC (continued)
  antitumor activity in
      animals (continued)
    L5178Y leukemia, 26
    P815 leukemia, 27
    sarcoma 180, 27
  apudomas, 115
  cervical cancer, 114
  clinical use
    combinations with other
        antitumor drugs, 134-
        135
    dose schedule, 133-134
    intra-arterial administra-
        tion, 135
  combination with immuno-
      stimulants, 139-140
    BCG, 139
    Corinebacterium parvum,
        139
    α-2-R-interferon, 139
    interleukin-2, 139
    tumor vaccines, 139
  Hodgkin's disease, 112
    drug combinations, 112
  malignant melanoma, 109-111
    four drug combinations,
        111
    metastatic to CNS, 135-136
    phase II study of combina-
        tions, 209
    three drug combinations,
        110
    two drug combinations, 110
  hydrolysis to 5-diazoimida-
      zole-4-carboxamide,
      15, 24
  and immunity in humans,
      138-140
    lymphocyte cytotoxicity,
        138
    lymphocyte subpopulations,
        138
  mechanism of action, 15
    DNA alkylation, 17
    O(6)-guanine alkylation,
        17
  metabolic activation, 15,
      19, 24
  metabolism, 15, 16, 19, 24
    N-demethylation, 15
    hydroxylation, 15
    and MTIC, 15
  pediatric malignancies, 114
    drug cobinations, 114
  photo-decomposition, 119-120
  resistance, 27-28
    cross resistance to BIC,
        27-28
  soft tissue sarcomas, 113

222

DTIC (continued)
  soft tissue sarcomas (con-
      tinued)
    drug combinations, 113-114
  synthesis, 25
  toxicity
    gastrointestinal, 136
    hepatic, 136-138

Ethazolastone
  hydrolysis, 102
  mechanism of action, 101-105
  DNA alkylation, 102-105
  metabolism, 101-102

Hemostasis
  1-aryl-3,3-dimethyltria-
      zenes, 49-50
  DTIC, 49-50
Herbicidal activity
  1-aryl-3,3-dimethyltria-
      zenes, 23
Heterocyclic triazenes, 28
  pyrazole derivatives, 28
  triazole derivatives, 28
Hexamethylmelamine
  antitumor activity in ani-
      mals, 173-175
    ADJ/PC6 plasmocytoma, 175
  mechanism of action, 175
  metabolism, 175
    hydroxylation, 175
    N-demethlylation, 175
  ovarian cancer, 173-174
    in drug combinations,
        173-174
Hodgkin's disease
  DTIC, 112
  DTIC in drug combinations,
      112
Hydrolysis
  acetoxymethyltriazenes, 8, 11
  1-aryl-3,3-dimethyltria-
      zenes, 29, 30, 47-48
  1-aryl-3-methyltriazenes, 1,
      8
  DTIC, 15, 24
  ethazolastone, 102
  hydroxymethyltriazenes, 7
  imidazotetrazinones, 93-95
  mitozolomide, 93
  temozolomide, 93
Hydroxylation
  1-aryl-3,3-dimethyltria-
      zenes, 35
  DTIC, 15
  hexamethylmelamines, 175
  N-methylmelamines, 180-181
  pentamethylmelamine, 176

Mechanism of action (contin-
    ued)
  pentamethylmelamine (contin-
      ued)
    hydroxylation, 176
  temozolomide, 19, 150-156
    DNA alkylation, 19, 102-
        105, 151-156
    DNA hypomethylation, 19-20
    DNA major groove, 152-154
    N(7)-guanine alkylation,
        151
    O(6)-guanine alkylation,
        19, 151
Melanoma (malignant)
  DTIC, 109, 111, 135-136
  DTIC in drug combinations,
      110-111, 135-136
  phase II study, 209
  mitozolomide, 187
Melanoma (malignant) metastat-
    ic to CNS, 135-136
Metabolism
  1-(4-acetylphenyl)-3,3-
      diethyltriazene, 100-
      101
  1-(4-acetylphenyl)-3,3-
      dimethyltriazene,
      100-101
  1-aryl-3,3-dimethyltria-
      zenes, 91-93
    antimetastatic action,
        47-49
    antitumor action, 31, 35
    conversion to 1-aryl-3-
        methyltriazens, 30, 35
    hydrolysis to diazonium,
        30
    N-demethylation, 35
    hydroxylation, 35
  1-aryl-3-methyltriazene
      production, 91-92
    formaldehyde production,
        91
    and hydroxymethyltria-
        zenes, 1, 93
    oxidation, 92-93
  5-diazoimidazole-4-carboxa-
      mide
    conversion to 2-azahypox-
        anthine, 24
  DM-COOK, 100, 101, 121
    conjugation, 121, 125-127
    N-demethylation, 121, 125-
        127
  DTIC, 15, 16, 19, 24, 93
    AIC excretion, 93
    conversion to MTIC, 15
    N-demethylation, 15

223

Mitozoloide (continued)
  mechanism of action (continued)
    O(6)-guanine alkylation,
      19, 150
  phase I studies, 187
  phase II studies, 187-188
    malignant melanoma, 187
    small cell carcinoma of
      the lung, 187
  structure-activity relationships, 149-150
  synthesis, 146
  toxicity, 187-188
MM-COOK
  antimetastatic activity, 208
    Lewis lung carcinoma, 208
    MCa mammary carcinoma, 208
  antitumor activity in animals, 99
    L1210 leukemia, 99

Ovarian cancer
  hexamethylmelamine, 173-174
  hexamethylmelamine in drug
    combinations, 173-174
  trimelamol, 177

Pediatric malignancies
  DTIC, 114
  DTIC in drug combinations,
    114
Pentamethylmelamine, 175-176
  antitumor activity, 175
  metabolism, 176
    N-demethylation, 176
    hydroxylation, 176
  toxicity, 175-176
Pharmacokinetics
  CB10-277, 121, 125-127
  imidazotetrazinones, 187-191
  temozolomide, 188-191
  trimelamol, 177-178
Phase I trials
  CB10-277, 127-129
  imidazotetrazinones, 187-191
  mitozolomide, 187
  temozolomide, 188, 191
  trimelamol, 177
Phase II trials
  DTIC in drug combinations,
    209
  imidazotetrazinones, 187, 191
  mitozolomide, 187-188
  trimelamol, 178-179
Photo decomposition
  DTIC, 119-120
Prodrugs (see also ethazolastone)

Metabolism (continued)
  DTIC (continued)
    hydrolysis to 5-diazo-
      imidazole-4-carboxa-
      mide, 15, 24
    hydroxylation, 15
    hydroxymethyl derivative
      excretion, 93
    metabolic activation, 15,
      19, 24
    photo decomposition, 119-
      120
  ethazolastone, 101-102
  imidazotetrazinones, 93-95
    analogs, 94-95
    hydrolysis, 93-95
  mitozolomide, 93-95
    hydrolysis, 93
    MTIC generation,
  pentamethylmelamine, 176
    N-demethylation, 176
    hydroxylation, 176
  temozolomide, 93-95, 101-102
    hydrolysis, 93,
    MTIC generation, 93
Metastatic potential
  DM-COOK, 52-54, 208
  DTIC, 52-54, 208
Methylation, see alkylation
N-Methylmelamines, 173-181
    (see also hexamethyl-
    melamine, pentamethyl-
    melamine, trimelamol)
  mechanism of action, 180-181
    DNA damage, 180-181
    N-demethylation, 180-181
    hydroxylation, 180-181
1-p-(3-methyl-1-triazeno)
      benzoic acid potassium
      salt, see MM-COOK
5-(3-Methyl-1-triazeno)imida-
      zole-4-carboxamide,
      see MTIC
Mitozolomide
  antitumor activity in ani-
      mals, 19, 186-187
  decomposition, 18-19, 147-
      149
  mechanism of action, 19,
      150-156
    cross resistance with
      chloroethylnitrosoureas,
      151
    DNA alkylation, 19, 151-
      156
    DNA hypomethylation, 19-20
    DNA major groove, 152-154
    N(7)-guanine alkylation,
      150